Quantum Information and Symmetry

Quantum Information and Symmetry

Special Issue Editors

Wiesław Leoński
Joanna K. Kalaga
Radosław Szczęśniak

MDPI • Basel • Beijing • Wuhan • Barcelona • Belgrade • Manchester • Tokyo • Cluj • Tianjin

Special Issue Editors

Wiesław Leoński
University of Zielona Góra
Poland

Joanna K. Kalaga
University of Zielona Góra
Poland

Radosław Szczęśniak
Częstochowa University of Technology
Poland

Editorial Office
MDPI
St. Alban-Anlage 66
4052 Basel, Switzerland

This is a reprint of articles from the Special Issue published online in the open access journal *Symmetry* (ISSN 2073-8994) (available at: https://www.mdpi.com/journal/symmetry/special_issues/Quantum_Information_Symmetry).

For citation purposes, cite each article independently as indicated on the article page online and as indicated below:

LastName, A.A.; LastName, B.B.; LastName, C.C. Article Title. *Journal Name* **Year**, *Article Number*, Page Range.

ISBN 978-3-03928-800-7 (Pbk)
ISBN 978-3-03928-801-4 (PDF)

© 2020 by the authors. Articles in this book are Open Access and distributed under the Creative Commons Attribution (CC BY) license, which allows users to download, copy and build upon published articles, as long as the author and publisher are properly credited, which ensures maximum dissemination and a wider impact of our publications.

The book as a whole is distributed by MDPI under the terms and conditions of the Creative Commons license CC BY-NC-ND.

Contents

About the Special Issue Editors vii

Preface to "Quantum Information and Symmetry" ix

Alexey N. Pyrkov, Tim Byrnes and Valentin V. Cherny
Solitonic Fixed Point Attractors in the Complex Ginzburg–Landau Equation for Associative Memories
Reprinted from: *Symmetry* 2020, 12, 24, doi:10.3390/sym12010024 1

Jan Peřina Jr. and Antonín Lukš
Quantum Behavior of a $\mathcal{PT}$-Symmetric Two-Mode System with Cross-Kerr Nonlinearity
Reprinted from: *Symmetry* 2019, 11, 1020, doi:10.3390/sym11081020 11

Kamila A. Szewczyk, Ewa A. Drzazga-Szczęśniak, Marcin W. Jarosik, Klaudia M. Szczęśniak and Sandra M. Binek
Characteristics of the s–Wave Symmetry Superconducting State in the $BaGe_3$ Compound
Reprinted from: *Symmetry* 2019, 11, 977, doi:10.3390/sym11080977 23

Joanna K. Kalaga
The Entanglement Generation in $\mathcal{PT}$-Symmetric Optical Quadrimer System
Reprinted from: *Symmetry* 2019, 11, 1110, doi:10.3390/sym11091110 35

Marco Enríquez, Alfonso Jaimes-Nájera and Francisco Delgado
Single-Qubit Driving Fields and Mathieu Functions
Reprinted from: *Symmetry* 2019, 11, 1172, doi:10.3390/sym11091172 45

Anna Kowalewska-Kudłaszyk and Grzegorz Chimczak
Asymmetry of Quantum Correlations Decay in Nonlinear Bosonic System
Reprinted from: *Symmetry* 2019, 11, 1023, doi:10.3390/sym11081023 61

Przemyslaw Tarasewicz
The Symmetry of Pairing and the Electromagnetic Properties of a Superconductor with a Four-Fermion Attraction at Zero Temperature
Reprinted from: *Symmetry* 2019, 11, 1358, doi:10.3390/sym11111358 71

About the Special Issue Editors

Wiesław Leoński (Prof. Dr hab.) works as a full professor at the Institute of Physics, University of Zielona Góra, Poland, where he is head of the Quantum Optics and Engineering Division. He also leads the newly established Laboratory of Quantum Optics and Optical Technologies. W. Leoński completed his studies at Adam Mickiewicz University in Poznań, Poland, and Keble College, University of Oxford, U.K. He gained a Ph.D. and postdoctoral degree at Adam Mickiewicz University. In 2015, the President of the Polish Republic conferred him the title of Professor in Physics. His research interests are related to quantum and nonlinear optics, quantum correlations, classical and quantum chaos, and quantum information theory. He has published over 100 scientific articles in international journals and presented over 100 conference communications. He is a member of the Polish Physical Society, European Optical Society, American Physical Society, Optical Society of America, and The American Association for the Advancement of Science.

Joanna K. Kalaga (Dr.) holds a Ph.D. in Physics from the University of Zielona Góra (Poland, 2012). She is currently working as an Assistant Professor in the Quantum Optics and Engineering Division at the Institute of Physics at the University of Zielona Góra. Her main research interests include quantum optics, quantum information theory, and quantum chaos. She has published 27 research articles (22 of them are indexed in Clarivate Analytics Web of Science) and a book chapter. Dr. Kalaga is a member of the European Optical Society (EOS).

Radosław Szczęśniak (Prof. Dr hab.). In 2002, Radosław Szczęśniak obtained a Ph.D. in Physics from the University of Silesia in Katowice, Poland. In 2013, he received thae postdoctoral degree awarded by the Wrocław University of Technology, Poland. Radosław Szczesniak works as a professor at the Częstochowa University of Technology. He conducts research in the fields of superconductivity, classical and quantum chaos, the theory of open systems, and quantum information.

Preface to "Quantum Information and Symmetry"

Quantum information theory (QIT) and its applications have become one of the most relevant fields in contemporary research related to both physical and technical sciences. We have approached the point in which novel quantum technologies are starting to play a crucial role in our world. Due to achievements in the field of QIT and its applications, we are living in the second quantum revolution era. As the broadly understood symmetries play a significant role in physics, they will also point to new research directions in the field of QIT.

The Special Issue is devoted to this overlapping of the ideas of QIT and methods and models in which various kinds of symmetries are applied. The published articles not only focus on theoretical concepts related to information theory but also discuss physical models that could be implemented in future designs of QIT devices. These devices could be built, for instance, as an implementation of models describing superconductivity phenomena, which are also discussed in the Special Issue. The broad range of other topics related to the symmetries and quantum information are presented and discussed. For instance, quantum correlations (including quantum entanglement), solitonic attractors, single-quit radiation fields, models of bosonic systems, PT-symmetry, Kerr-type nonlinearities, and others are the subjects of the articles presented in the Issue.

Wiesław Leoński, Joanna K. Kalaga, Radosław Szczęśniak
Special Issue Editors

Article

Solitonic Fixed Point Attractors in the Complex Ginzburg–Landau Equation for Associative Memories

Alexey N. Pyrkov [1,*], Tim Byrnes [2,3,4,5,6] and Valentin V. Cherny [1]

1. Institute of Problems of Chemical Physics of Russian Academy of Sciences, Acad. Semenov av. 1, Chernogolovka, 142432 Moscow, Russia; vcherny2009@gmail.com
2. Department of Physics, New York University Shanghai, 1555 Century Ave, Pudong, Shanghai 200122, China; tim.byrnes@nyu.edu
3. State Key Laboratory of Precision Spectroscopy, School of Physical and Material Sciences, East China Normal University, Shanghai 200062, China
4. NYU-ECNU Institute of Physics at NYU Shanghai, 3663 Zhongshan Road North, Shanghai 200062, China
5. National Institute of Informatics, 2-1-2 Hitotsubashi, Chiyoda-ku, Tokyo 101-8430, Japan
6. Department of Physics, New York University, New York, NY 10003, USA
* Correspondence: alex.pyrkov@gmail.com

Received: 27 November 2019; Accepted: 16 December 2019; Published: 20 December 2019

Abstract: It was recently shown that the nonlinear Schrodinger equation with a simplified dissipative perturbation features a zero-velocity solitonic solution of non-zero amplitude which can be used in analogy to attractors of Hopfield's associative memory. In this work, we consider a more complex dissipative perturbation adding the effect of two-photon absorption and the quintic gain/loss effects that yields the complex Ginzburg–Landau equation (CGLE). We construct a perturbation theory for the CGLE with a small dissipative perturbation, define the behavior of the solitonic solutions with parameters of the system and compare the solution with numerical simulations of the CGLE. We show, in a similar way to the nonlinear Schrodinger equation with a simplified dissipation term, a zero-velocity solitonic solution of non-zero amplitude appears as an attractor for the CGLE. In this case, the amplitude and velocity of the solitonic fixed point attractor does not depend on the quintic gain/loss effects. Furthermore, the effect of two-photon absorption leads to an increase in the strength of the solitonic fixed point attractor.

Keywords: attractor; complex Ginzburg–Landau equation; soliton; quantum machine learning; associative memory

---

1. Introduction

Neuromorphic computing—the study of information processing using articficial systems mimicking neuro-biological architectures—has attracted a huge amount of interest in modern information science [1–5]. With the recent explosion of interest in quantum information processing systems, it is of great interest whether neuromorphic computing can be combined with quantum approaches [6–8]. One of the best-known model systems in neuromorphic computing is the Hopfield's associative memory [9], which can be considered as a dissipative dynamical system with the ability to make associations [10–12]. In this case, the input state is one of stored patterns distorted by noise, and the convergence to the attractor can be understood as recognition of the distorted pattern. Hopfield's associative memory is usually applied to store finite dimensional vectors with dynamics described by a system of ordinary differential equations, which places restrictions on the patterns that can be processed. It has also been shown that it is possible to encode the information in attractors within a large dimensional dynamical system with functional configuration space. This allows for the storage and recovery of quite complex and strongly distorted data structures. However, for partial differential equations of a relatively general form, there are no algorithms for the determination of the desired values of system parameters, which turn a given point of functional space into an attractor.

Meanwhile, it was shown that certain solitonic evolutionary partial differential equations, which admit solutions of the form of localized waves with complex topological structures, can be applied to machine learning [13]. Since such equations are usually conservative, their solutions necessarily describe the relative motion and interaction of a constant number of solitons, which are determined by the initial conditions. The nonlinear Schrodinger equation (NLSE) is one of the best known solitonic equations which has already been applied in very different fields of science. It provides impressively precise description of many physical systems, from vortex filaments to superfluids [14–16]. The Gross–Pitaevskii equation is one particular case of the NLSE and captures many aspects of the time evolution of Bose–Einstein condensates (BECs) [17–19]. An open dissipative version of the NLSE can also be realized in exciton–polariton BEC systems exhibiting solitons [20–24], optical fibers [25–27] and microresonators [28–30]. The use of BECs to solve classical optimization problems [31,32] and perform quantum algorithms [33–40] have also been investigated. Furthermore, the NLSE has a number of different symmetries [41], and there has been much interest in linear and nonlinear properties of systems with potentials obeying parity-time ($\mathcal{PT}$) symmetry [42–45]. Experimental observation of $\mathcal{PT}$ symmetry breaking in optics, and several theoretical suggestions of realization of $\mathcal{PT}$-symmetric optical systems have been made in recent years [46–51].

Recently, the nonlinear Schrodinger equation (NLSE), which can be realized in BEC, with a simplified dissipative perturbation which creates a frictional force acting on soliton [25,52–56] was considered in an application to associative memory and pattern recognition [57]. It was shown that the control of the perturbative term allows one to decrease the velocity of soliton to zero and conserve a positive value of its amplitude. The perturbation makes the zero-velocity solitonic solution of non-zero amplitude into an attractor for all evolution trajectories whose initial conditions are moving solitons. This paves the way to store information in a large dimensional dynamical system using principles which are completely analogous to that of Hopfield's associative memory.

In this paper, we consider the complex Ginzburg–Landau equation (CGLE) [41,58] and show that it has similar properties as seen in Ref. [57] that can be exploited towards associative memory and pattern recognition. The CGLE is of interest since it can be implemented in experimentally accessible systems such as nonlinear optics, which can form the basis of experimental realization of the general approach. We construct a perturbation theory for CGLE and compare the solution with numerical simulations. We show that similarly to the simplified model, a zero-velocity solitonic solution of non-zero amplitude appears in the CGLE, and we investigate the behavior of the solitonic solution on various parameter choices.

2. The Complex Ginzburg–Landau Equation

We consider the CGLE with a dissipative perturbation which creates a frictional force on the soliton. We show that control of the term allows us to decrease the velocity of the soliton to zero and retain some positive value of its amplitude. The existence of such a frictional force would mean that the perturbation turns the resting solitonic solution of certain amplitude into an attractor for all evolution trajectories with initial conditions that are moving solitons.

The CGLE with a small dissipative perturbation reads

$$iu_t + u_{xx} + 2|u|^2 u = \epsilon(iAu_{xx} + iBu + iC|u|^2 u + D|u|^4 u), \tag{1}$$

where the subscripts denote derivatives with respect to the variable, $0 < \epsilon \ll 1$ is a small parameter characterizing the perturbation, and A, B, C, D are real positive constants. The fundamental monosolitonic solution of LHS of Equation (1) are

$$f(x,t) = a\mathrm{sech}(a(x - vt - x_0))e^{i\frac{1}{2}vx + (a^2 - \frac{1}{4}v^2)t - i\sigma_0}, \tag{2}$$

where a is the soliton amplitude, v is its velocity, and x_0 with σ_0 are determined by the initial position and phase of the pulse. From Equation (2) we can see that the left hand side (LHS) of Equation (1) admits moving and steady state solitonic solutions. The small conservative perturbation of the right hand side (RHS) of Equation (1) makes the soliton oscillate around the minimum of this perturbation potential in a similar way to a classical particle.

From the results obtained in Ref. [59], it is natural to expect that the first term on the RHS of (1) will create a viscous friction force that will slow down and eventually stop the soliton. However, it also would not be surprising for such a frictional force to make the amplitude decay as well. The second dissipative term is known to increase the soliton amplitude without making changes in its velocity. The third and fourth terms describe the effect of two-photon absorption and the quintic gain/loss effects respectively.

One possible way to realize the dissipation is to use solitons in a BEC. If the confining potential, which stabilizes a BEC with the attractive interactions against collapse, is made asymmetric such that the atoms can only undergo one-dimensional (1D) motion, it has been predicted to have matter wave soliton solutions [60,61]. For an atomic BEC, the sign and magnitude of the nonlinearity is determined by the scattering length α. The interactions are repulsive for $\alpha > 0$ and attractive for $\alpha < 0$. Dark solitons have been observed in BECs for repulsive interactions [62–64]. With the use of Feshbach resonances [65], adjusting the atom–atom interaction from repulsive to attractive, bright matter-wave solitons and soliton trains were created in a BEC [66,67]. For exciton-polaritons, the interactions can be tuned by changing the exciton fraction or introducing an indirect exciton component which possess dipolar forces [68,69]. Furthermore, it was shown that matter–wave bright solitons can form entangled states [70].

In an implementation with a fiber laser the dissipation can be produced in the following way. The combined effects of self-phase modulation and cross-phase modulation induced on two orthogonal polarization components produces a non-linearity during the propagation of the pulse in the fiber. A polarization controller is adjusted at the output of the fiber such that the polarizing isolator passes the central intense part of the pulse but blocks the low-intensity pulse wings. In the regime where low-intensity waves are not as efficiently filtered out, the existence of a continuous wave (cw) component that mediates interactions between solitons strongly affect the dynamics and a large number of quasi-cw components produce a noisy background from which dissipative solitons can be formed in the fiber laser cavity and reach the condensed phase. The soliton flow can be adjusted by manual cavity tuning or triggered by the injection of an external low-power cw laser [25,54,71–76].

3. Solitonic Fixed Point Attractors

We apply Lagrangian perturbation theory for conservative partial differential equations to describe the soliton behavior under the chosen perturbation. In this case we assume that the solution of the perturbed equation with a single soliton initial condition continues to have this form under evolution but the four characterizing parameters become time-dependent. This assumption is valid for sufficiently small values of ϵ. Thus, we can rewrite this solution in the following form

$$u(x,t) = a\operatorname{sech}(a\theta)e^{i\zeta\theta+i\sigma}, \qquad (3)$$

where $\theta = x - 2\zeta t - x_0$, $\zeta = \frac{v}{2}$, $\sigma = (a^2 + \zeta^2)t - \sigma_0 + \zeta x_0$. The LHS of Equation (1) describes a conservative complex scalar field and can thus be (along with its complex conjugate equation) derived from the Lagrangian density

$$\mathcal{L} = \frac{i}{2}(u^* u_t - u_t^* u) - |u_x|^2 + |u|^4, \qquad (4)$$

where u^* denotes the complex conjugate to the u field variable. The field u and its complex conjugate u^* can be considered independent fields. For this reason they can be taken as the generalized Lagrangian coordinates of our problem. By making a variation of $\mathcal{L}$ by u^* and u we can obtain the corresponding densities of generalized momentum with conventional expressions from Hamiltonian mechanics:

$$\pi = \frac{\partial \mathcal{L}}{\partial u_t} = +\frac{i}{2}u^* \qquad (5)$$

$$\pi^* = \frac{\partial \mathcal{L}}{\partial u_t^*} = -\frac{i}{2}u, \qquad (6)$$

such that the Hamiltonian density takes the form

$$\mathfrak{H} = u_t \pi + u_t^* \pi^* - \mathcal{L} = |u_x|^2 - |u|^4. \qquad (7)$$

The complete energy functional is then given by

$$H = \int_{-\infty}^{\infty} \mathfrak{H} dx. \qquad (8)$$

Then we can obtain the following equation

$$\frac{dH}{dt} = -\int_{-\infty}^{\infty} (u_{xx} + 2|u|^2 u) u_t^* dx + \text{c.c.}$$

$$= -\epsilon \int_{-\infty}^{\infty} iR u_t^* dx + \text{c.c.}, \qquad (9)$$

where c.c. denotes the complex conjugate expression and R is a perturbation term of a general form.

To apply perturbation theory to our case, we assume parameters of the soliton to be time-independent, and calculate the complete field Lagrangian for the single-soliton initial condition under this assumption. This can be done through direct substitution of (3) into the Lagrangian density (4) with subsequent integration over the coordinate space:

$$L = \int_{-\infty}^{\infty} \mathcal{L} dx + \text{c.c.} \qquad (10)$$

This procedure gives the following expression for the Lagrangian in terms of soliton parameters:

$$L = \frac{2}{3}a^3 - 2a\xi^2 + 2a\xi\frac{d\alpha}{dt} - 2a\frac{d\sigma}{dt}, \qquad (11)$$

where $\alpha = x - \theta = 2\xi t + x_0$ is the fourth independent parameter of the soliton. It is now convenient to rewrite these parameters as a four-dimensional tuple

$$\{a, \xi, \alpha, \sigma\} = \{y_1(t), y_2(t), y_3(t), y_4(t)\}. \qquad (12)$$

In this notation, the main equation in our case has the following form [77]:

$$\frac{\partial L}{\partial y_i} - \frac{d}{dt}\left(\frac{\partial L}{\partial y_{i,t}}\right) = \epsilon \int_{-\infty}^{\infty} iR \frac{\partial u^*}{\partial y_i} dx + \text{c.c.} \qquad (13)$$

Thus for $R = A u_{xx} + B u + C|u|^2 u - iD|u|^4 u$, we can obtain the system of ordinary differential equations for the parameters of the perturbed soliton

$$\dot{\sigma} = a^2(1 - \frac{8\epsilon D a^2}{9}) + \check{\varsigma}^2 \qquad (14)$$

$$\dot{\alpha} = 2\check{\varsigma} \qquad (15)$$

$$\dot{a} = \epsilon(\frac{-2A}{3}a^3 - 2Aa\check{\varsigma}^2 + \frac{4C}{3}a^3 + 2Ba) \qquad (16)$$

$$\dot{\check{\varsigma}} = \frac{-4\epsilon A a^2 \check{\varsigma}}{3}. \qquad (17)$$

First of all, we analyze the last equation, which determine the evolution of the velocity $\check{\varsigma}$. Since both $\check{\varsigma}$ and a are always non-negative, the RHS of this equation is negative so the velocity decays with time. Assuming that there exists a stationary point, we can identify the left side of the equation with zero, thus with the condition of $a \neq 0$ we obtain

$$\dot{\check{\varsigma}} = 0 \Rightarrow \frac{-4\epsilon A a^2 \check{\varsigma}}{3} = 0 \Rightarrow \check{\varsigma} = 0. \qquad (18)$$

That is, there is only one stationary point of zero velocity as required.

Next consider (16) for the amplitude a. According to above assumptions, the first two terms on the RHS of this equation decrease the amplitude, while the last two increase it. The attainability of the equilibrium between those forces would mean the existence of an attractive stationary point. We assume that such point exists for $t^* \gg 1$, and the following relation holds: $\dot{a}(t \geq t^*) \approx 0$. It then follows that

$$\dot{a} = 0 \Rightarrow \frac{-Aa^3}{3} - Aa\check{\varsigma}^2 + \frac{2Ca^3}{3} + Ba = 0$$

$$\Rightarrow a = \sqrt{\frac{3(B - A\check{\varsigma}^2)}{A - 2C}} \qquad (19)$$

and for times large enough $\check{\varsigma} = 0$ the expression for a^* is

$$a^* = \sqrt{\frac{3B}{A - 2C}}. \qquad (20)$$

This means, that if $A - 2C > 0$ and $B \neq 0$ our system has an attractor in a form of soliton with positive amplitude a^* whose velocity equals to zero. From the expressions we can see that velocity and amplitude of the attractor do not depend on the quintic gain/loss effects. At the same time, increasing of C leads to faster slowing down the soliton and increasing of the attractors amplitude.

Figure 1 shows a numerical evolution of the derived ODE system describing the perturbation theory approximation for different parameters A, B, C. We can see that in this case a standing state soliton at the minimum of the potential part of the perturbation is translated into a attractor for any monosolitonic initial condition and that for any given values of A, B, C the ODE system has only one attractor. Since the expression for a^* depends only on the $\frac{B}{A-2C}$, time dependences of soliton amplitude and velocity are presented on Figure 1a,b for different choices of parameter A and fixed B, C. We see from Figure 1 that increasing A causes the velocity of soliton to decrease faster and that the amplitude of the soliton decreases with increasing A. Thus, by controlling the parameters A, B, C we can control the amplitude and velocity of soliton towards an attractor that can be designed to store and restore information. Figure 2 shows a simulation of the CGLE in the form (1) together with numerical evaluation of the derived ODE system describing the perturbation theory approximation. We can see that predictions of perturbation theory are in good agreement with the numerical solution. All calculations are implemented for the optimal parameter of $\epsilon = 0.001$. For larger values of ϵ we have faster convergence to the fixed point solution, but in this case the solution becomes unstable for longer times (for instance, at $\epsilon = 0.005$ the solitonic solution converges to fixed point solution at about

$t \approx 200$ but becomes unstable for $t > 600$). On the other hand, for smaller values of ϵ the solution is stable everywhere under consideration but it converges to the fixed point solution very slowly.

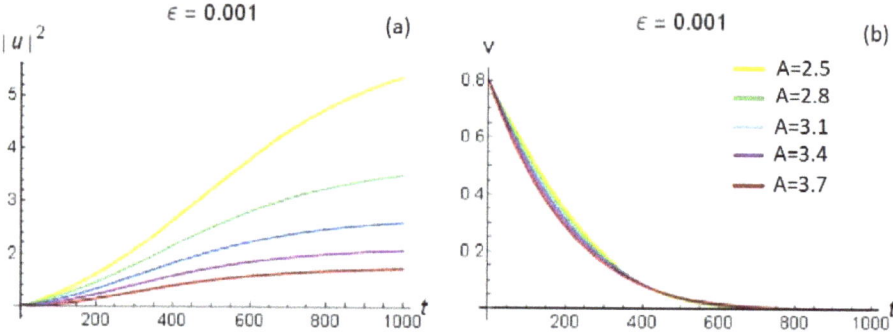

Figure 1. (a) Amplitude and (b) velocity of the soliton versus different A for $B = C = 1$.

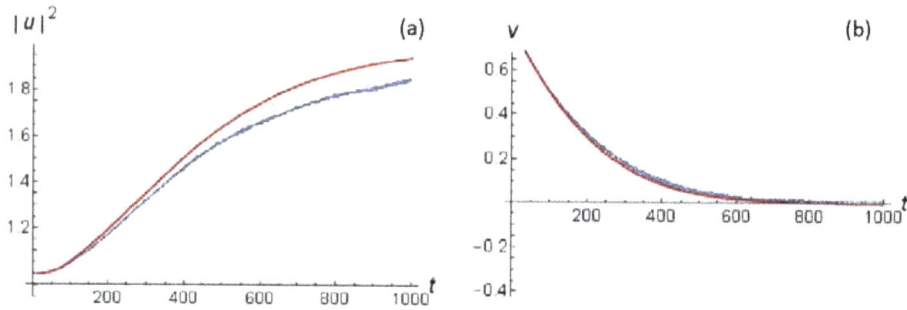

Figure 2. Evolution of soliton amplitude (a) and velocity (b) with time. Comparison of perturbation theory (red solid line) with numerical solution (blue dot line) for $\epsilon = 0.001$. Parameters of initial soliton are $a = 1, \zeta = \frac{4}{5}$ and dissipative parameters are $A = 3.5, B = C = 1$.

4. Conclusions

We have investigated the CGLE with a small dissipation term, both analytically and numerically, to realize solitonic fixed point attractors. We have shown that, in this case, a standing state soliton in the minimum of potential part of perturbation is translated into an attractor for any monosolitonic initial conditions. It is shown that the control of dissipative perturbation allows us to handle the attractor of system similarly to Ref. [57], such that it is possible to store and process information. This approach can be realized with solitons in Bose–Einstein condensates and nonlinear optical systems.

Author Contributions: A.N.P. and V.V.C. performed the calculations. All authors discussed the results and wrote the paper. All authors have read and agreed to the published version of the manuscript.

Funding: The work is supported by the RFBR-NSFC collaborative program (Grant No. 18-57-53007) and the State assignment (N. 0089-2019-0002). T.B. is supported by the Shanghai Research Challenge Fund; New York University Global Seed Grants for Collaborative Research; National Natural Science Foundation of China (61571301, D1210036A); the NSFC Research Fund for International Young Scientists (11650110425, 11850410426); NYU-ECNU Institute of Physics at NYU Shanghai; the Science and Technology Commission of Shanghai Municipality (17ZR1443600); the China Science and Technology Exchange Center (NGA-16-001); and the NSFC-RFBR Collaborative grant (81811530112).

Conflicts of Interest: The authors declare no conflict of interest.

References

1. Monroe, D. Neuromorphic Computing Gets Ready For the (Really) Big Time. *Commun. ACM* **2014**, *57*, 13–15. [CrossRef]
2. Zhao, W.S.; Agnus, G.; Derycke, V.; Filoramo, A.; Bourgoin, J.P.; Gamrat, C. Nanotube devices based crossbar architecture: Toward neuromorphic computing. *Nanotechnology* **2010**, *21*, 175202. [CrossRef] [PubMed]
3. Mead, C. Neuromorphic electronic systems. *Proc. IEEE* **1990**, *78*, 1629–1636. [CrossRef]
4. Sheridan, P.M.; Cai, F.; Du, C.; Zhang, Z.; Lu, W.D. Sparse coding with memristor networks. *Nat. Nanotechnol.* **2017**, *12*, 784–789. [CrossRef] [PubMed]
5. Sebastian, A.; Tuma, T.; Papandreou, N.; Le Gallo, M.; Kull, L.; Parnell, T.; Eleftheriou, E. Temporal correlation detection using computational phase-change memory. *Nat. Commun.* **2017**, *8*, 1115. [CrossRef] [PubMed]
6. Nielsen, M.A.; Chuang, I.L. *Quantum Computation and Quantum Information*; Cambridge University Press: Cambridge, UK, 2000.
7. Preskill, J. Quantum Computing in the NISQ era and beyond. *Quantum* **2018**, *2*, 79. [CrossRef]
8. Lamata, L.; Sanz, M.; Solano, E. Quantum Machine Learning and Bioinspired Quantum Technologies. *Adv. Quantum Technol.* **2019**, *2*, 1900075. [CrossRef]
9. Hopfield, J.J. Neural Networks and Physical Systems with Emergent Collective Computational Abilities. *Proc. Natl. Acad. Sci. USA* **1982**, *79*, 2554–2558. [CrossRef]
10. Izhikevich, E. *Dynamical Systems in Neuroscience: The Geometry of Excitability and Bursting*; The MIT Press: Cambridge, MA, USA, 2007.
11. Strogatz, S. *Nonlinear Dynamics and Chaos: With Applications to Physics, Biology and Chemistry*; Perseus: New York, NY, USA, 2001.
12. Hertz, J.; Krogh, A.; Palmer, R.G. *Introduction to the Theory of Neural Computation*; Addison-Wesley: Redwood City, CA, USA, 1991.
13. Behera, L.; Kar, I.; Elitzur, A. A Recurrent Quantum Neural Network Model to Describe Eye Tracking of Moving Targets. *Found. Phys. Lett.* **2005**, *18*, 357–370. [CrossRef]
14. Onorato, M.; Proment, D.; Clauss, G.; Klein, M. Rogue Waves: From Nonlinear Schrödinger Breather Solutions to Sea-Keeping Test. *PLoS ONE* **2013**, *8*, e54629. [CrossRef]
15. Pitaevskii, L.; Stringari, S. *Bose-Einstein Condensation*; Clarendon: Oxford, UK, 2003.
16. Falkovich, G. *Fluid Mechanics (A Short Course for Physicists)*; Cambridge University Press: Cambridge, UK, 2011.
17. Dalfovo, F.; Giorgini, S.; Pitaevskii, L.P.; Stringari, S. Theory of Bose-Einstein condensation in trapped gases. *Rev. Mod. Phys.* **1999**, *71*, 463–512. [CrossRef]
18. Bagnato, V.S.; Frantzeskakis, D.J.; Kevrekidis, P.G.; Malomed, B.A.; Mihalache, D. Bose-Einstein condensation: Twenty years after. *arXiv* **2015**, arXiv:1502.06328.
19. Campbell, R.; Oppo, G.-L. Stationary and traveling solitons via local dissipation in Bose-Einstein condensates in ring optical lattices. *Phys. Rev. A* **2016**, *94*, 043626. [CrossRef]
20. Byrnes, T.; Kim, N.Y.; Yamamoto, Y. Exciton–polariton condensates. *Nat. Phys.* **2014**, *10*, 803–813. [CrossRef]
21. Wouters, M.; Carusotto, I. Excitations in a nonequilibrium Bose-Einstein condensate of exciton polaritons. *Phys. Rev. Lett.* **2007**, *99*, 140402. [CrossRef] [PubMed]
22. Amo, A.; Pigeon, S.; Sanvitto, D.; Sala, V.; Hivet, R.; Carusotto, I.; Pisanello, F.; Leménager, G.; Houdré, R.; Giacobino, E.; et al. Polariton superfluids reveal quantum hydrodynamic solitons. *Science* **2011**, *332*, 1167–1170. [CrossRef]
23. Sich, M.; Krizhanovskii, D.; Skolnick, M.; Gorbach, A.V.; Hartley, R.; Skryabin, D.V.; Cerda-Méndez, E.; Biermann, K.; Hey, R.; Santos, P. Observation of bright polariton solitons in a semiconductor microcavity. *Nat. Photonics* **2012**, *6*, 50. [CrossRef]
24. Egorov, O.; Skryabin, D.V.; Yulin, A.; Lederer, F. Bright cavity polariton solitons. *Phys. Rev. Lett.* **2009**, *102*, 153904. [CrossRef]
25. Grelu, P.; Akhmediev, N. Dissipative solitons for mode-locked lasers. *Nat. Photonics* **2012**, *6*, 84–92. [CrossRef]
26. Wright, L.G.; Christodoulides, D.N.; Wise, F.W. Spatiotemporal mode-locking in multimode fiber lasers. *Science* **2017**, *358*, 94–97. [CrossRef]

27. Gustave, F.; Radwell, N.; McIntyre, C.; Toomey, J.P.; Kane, D.M.; Barland, S.; Firth, V.J.; Oppo, G.-L.; Ackemann, T. Observation of Mode-Locked Spatial Laser Solitons. *Phys. Rev. Lett.* **2017**, *118*, 044102. [CrossRef]
28. Kippenberg, T.J.; Gaeta, A.L.; Lipson, M.; Gorodetsky, M.L. Dissipative Kerr solitons in optical microresonators. *Science* **2018**, *361*, eaan8083. [CrossRef]
29. Suh, M.G.; Yang, Q.F.; Yang, K.Y.; Yi, X.; Vahala, K.J. Microresonator soliton dual-comb spectroscopy. *Science* **2016**, *354*, 600–603. [CrossRef] [PubMed]
30. Stone, J.R.; Briles, T.C.; Drake, T.E.; Spencer, D.T.; Carlson, D.R.; Diddams, S.A.; Papp, S.B. Thermal and Nonlinear Dissipative-Soliton Dynamics in Kerr-Microresonator Frequency Combs. *Phys. Rev. Lett.* **2018**, *121*. [CrossRef] [PubMed]
31. Byrnes, T.; Yan, K.; Yamamoto, Y. Accelerated optimization problem search using Bose-Einstein condensation. *New J. Phys.* **2011**, *13*, 113025. [CrossRef]
32. Byrnes, T.; Koyama, S.; Yan, K.; Yamamoto, Y. Neural networks using two-component Bose-Einstein condensates. *Sci. Rep.* **2013**, *3*, 2531. [CrossRef] [PubMed]
33. Byrnes, T.; Rosseau, D.; Khosla, M.; Pyrkov, A.; Thomasen, A.; Mukai, T.; Koyama, S.; Abdelrahman, A.; Ilo-Okeke, E. Macroscopic quantum information processing using spin coherent states. *Opt. Commun.* **2015**, *337*, 102–109. [CrossRef]
34. Pyrkov, A.N.; Byrnes, T. Entanglement generation in quantum networks of Bose-Einstein condensates. *New J. Phys.* **2013**, *15*, 093019. [CrossRef]
35. Pyrkov, A.N.; Byrnes, T. Full-Bloch-sphere teleportation of spinor Bose-Einstein condensates and spin ensembles. *Phys. Rev. A* **2014**, *90*, 062336. [CrossRef]
36. Byrnes, T.; Wen, K.; Yamamoto, Y. Macroscopic quantum computation using Bose-Einstein condensates. *Phys. Rev. A* **2012**, *85*, 040306. [CrossRef]
37. Gross, C. Spin squeezing, entanglement and quantum metrology with Bose-Einstein condensates. *J. Phys. B At. Mol. Phys.* **2012**. [CrossRef]
38. Pyrkov, A.N.; Byrnes, T. Quantum information transfer between two-component Bose-Einstein condensates connected by optical fiber. *Proc. SPIE* **2013**, *8700*, 87001E. [CrossRef]
39. Pyrkov, A.N.; Byrnes, T. Quantum information processing with macroscopic two-component Bose-Einstein condensates. In Proceedings of the International Conference on Micro- and Nano-Electronics 2018, Zvenigorod, Russia, 1–5 October 2018. [CrossRef]
40. Hecht, T. Quantum Computation with Bose-Einstein Condensates. Master's Thesis, Technische Universität München, München, Germany, 2004.
41. Aranson, I.S.; Kramer, L. The world of the complex Ginzburg-Landau equation. *Rev. Mod. Phys.* **2002**, *74*, 99–143. [CrossRef]
42. Bender, C.M.; Boettcher, S. Real Spectra in Non-Hermitian Hamiltonians Having PT-Symmetry. *Phys. Rev. Lett.* **1998**, *80*, 5243–5246. [CrossRef]
43. Kartashov, Y.V.; Malomed, B.A.; Torner, L. Unbreakable PT symmetry of solitons supported by inhomogeneous defocusing nonlinearity. *Opt. Lett.* **2014**, *39*, 5641. [CrossRef]
44. Chen, Y.; Yan, Z.; Mihalache, D.; Malomed, B.A. Families of stable solitons and excitations in the PT-symmetric nonlinear Schrödinger equations with position-dependent effective masses. *Sci. Rep.* **2017**, *7*, 1257. [CrossRef]
45. Bender, C.M. Making sense of non-Hermitian Hamiltonians. *Rep. Progress Phys.* **2007**, *70*, 947–1018. [CrossRef]
46. Longhi, S. Bloch Oscillations in Complex Crystals with PT Symmetry. *Phys. Rev. Lett.* **2009**, *103*, 123601. [CrossRef]
47. Guo, A. Observation of PT-Symmetry Breaking in Complex Optical Potentials. *Phys. Rev. Lett.* **2009**, *103*, 093902. [CrossRef]
48. Ruter, C.E.; Makris, K.G.; El-Ganainy, R.; Christodoulides, D.N.; Segev, M.; Kip, D. Observation of parity–time symmetry in optics. *Nat. Phys.* **2010**, *6*, 192. [CrossRef]
49. Regensburger, A.; Bersch, C.; Miri, M.A.; Onishchukov, G.; Christodoulides, D.N.; Peschel, U. Parity–time synthetic photonic lattices. *Nature* **2012**, *488*, 7410. [CrossRef] [PubMed]
50. Hodaei, H.; Miri, M.A.; Heinrich, M.; Christodoulides, D.N.; Khajavikhan, M. Parity-time-symmetric microring lasers. *Science* **2014**, *346*, 975. [CrossRef] [PubMed]

51. Wimmer, M.; Regensburger, A.; Miri, M.A.; Bersch, C.; Christodoulides, D.N.; Peschel, U. Observation of optical solitons in PT-symmetric lattices. *Nat. Commun.* **2015**, *6*, 8782. [CrossRef] [PubMed]
52. Kivshar, Y.S.; Malomed, B.A. Dynamics of solitons in nearly integrable systems. *Rev. Mod. Phys.* **1989**, *61*, 763–915. [CrossRef]
53. Malomed, B.A.; Mihalache, D.; Wise, F.; Torner, L. Spatiotemporal optical solitons. *J. Opt. B Quantum Semiclassical Opt.* **2005**, *7*, R53–R72. [CrossRef]
54. Mihalache, D. Multidimensional localized structures in optical and matter-wave media: A topical survey of recent literature. *Rom. Rep. Phys.* **2017**, *69*, 403.
55. Malomed, B.A. Evolution of nonsoliton and "quasi-classical" wavetrains in nonlinear Schrödinger and Korteweg-de Vries equations with dissipative perturbations. *Phys. D Nonlinear Phenom.* **1987**, *29*, 155–172. [CrossRef]
56. Malomed, B.A. Bound solitons in the nonlinear Schrödinger–Ginzburg-Landau equation. *Phys. Rev. A* **1991**, *44*, 6954–6957. [CrossRef]
57. Cherny, V.V.; Byrnes, T.; Pyrkov, A.N. Nontrivial Attractors of the Perturbed Nonlinear Schrödinger Equation: Applications to Associative Memory and Pattern Recognition. *Adv. Quantum Technol.* **2019**, *2*, 1800087. [CrossRef]
58. García-Morales, V.; Krischer, K. The complex Ginzburg-Landau equation: An introduction. *Contemp. Phys.* **2012**, *53*, 79–95. [CrossRef]
59. Leblond, H. Dissipative solitons: The finite bandwidth of gain as a viscous friction. *Phys. Rev. A* **2016**, *93*, 013830. [CrossRef]
60. Pérez-García, V.M.; Michinel, H.; Herrero, H. Bose-Einstein solitons in highly asymmetric traps. *Phys. Rev. A* **1998**, *57*, 3837–3842. [CrossRef]
61. Reinhardt, W.P.; Clark, C.W. Soliton dynamics in the collisions of Bose - Einstein condensates: an analogue of the Josephson effect. *J. Phys. B At. Mol. Phys.* **1997**, *30*, L785–L789. [CrossRef]
62. Burger, S.; Bongs, K.; Dettmer, S.; Ertmer, W.; Sengstock, K.; Sanpera, A.; Shlyapnikov, G.V.; Lewenstein, M. Dark Solitons in Bose-Einstein Condensates. *Phys. Rev. Lett.* **1999**, *83*, 5198–5201. [CrossRef]
63. Denschlag, J.; Simsarian, J.E.; Feder, D.L.; Clark, C.W.; Collins, L.A.; Cubizolles, J.; Deng, L.; Hagley, E.W.; Helmerson, K.; Reinhardt, W.P.; et al. Generating Solitons by Phase Engineering of a Bose-Einstein Condensate. *Science* **2000**, *287*, 97–101. [CrossRef]
64. Dutton, Z.; Budde, M.; Slowe, C.; Vestergaard Hau, L. Observation of Quantum Shock Waves Created with Ultra- Compressed Slow Light Pulses in a Bose-Einstein Condensate. *Science* **2001**, *293*, 663–668. [CrossRef]
65. Chin, C.; Grimm, R.; Julienne, P.; Tiesinga, E. Feshbach resonances in ultracold gases. *Rev. Mod. Phys.* **2010**, *82*, 1225–1286. [CrossRef]
66. Khaykovich, L.; Schreck, F.; Ferrari, G.; Bourdel, T.; Cubizolles, J.; Carr, L.D.; Castin, Y.; Salomon, C. Formation of a Matter-Wave Bright Soliton. *Science* **2002**, *296*, 1290–1293. [CrossRef]
67. Strecker, K.E.; Partridge, G.B.; Truscott, A.G.; Hulet, R.G. Formation and propagation of matter-wave soliton trains. *Nature* **2002**, *417*, 150–153. [CrossRef]
68. Byrnes, T.; Recher, P.; Yamamoto, Y. Mott transitions of exciton polaritons and indirect excitons in a periodic potential. *Phys. Rev. B* **2010**, *81*, 205312. [CrossRef]
69. Byrnes, T.; Kolmakov, G.V.; Kezerashvili, R.Y.; Yamamoto, Y. Effective interaction and condensation of dipolaritons in coupled quantum wells. *Phys. Rev. B* **2014**, *90*, 125314. [CrossRef]
70. Tsarev, D.V.; Arakelian, S.M.; Ray-Kuang Lee, Y.L.C.; Alodjants, A.P. Quantum metrology beyond Heisenberg limit with entangled matter wave solitons. *Opt. Express* **2018**, *26*, 19583–19595. [CrossRef] [PubMed]
71. Malomed, B.A. Multidimensional solitons: Well-established results and novel findings. *Eur. Phys. J. Spec. Top.* **2016**, *225*. [CrossRef]
72. Soto-Crespo, J.M.; Akhmediev, N.; Grelu, P.; Belhache, F. Quantized separations of phase-locked soliton pairs in fiber lasers. *Opt. Lett.* **2003**, *28*, 1757–1759. [CrossRef]
73. Komarov, A.; Komarov, K.; Leblond, H.; Sanchez, F. Spectral-selective management of dissipative solitons in passive mode-locked fibre lasers. *J. Opt. A Pure Appl. Opt.* **2007**, *9*, 1149–1156. [CrossRef]
74. Tang, D.Y.; Man, W.S.; Tam, H.Y.; Drummond, P.D. Observation of bound states of solitons in a passively mode-locked fiber laser. *Phys. Rev. A* **2001**, *64*, 033814. [CrossRef]
75. Chouli, S.; Grelu, P. Rains of solitons in a fiber laser. *Opt. Express* **2009**, *17*, 11776. [CrossRef]

76. Ben Braham, F.; Semaan, G.; Bahloul, F.; Salhi, M.; Sanchez, F. Experimental optimization of dissipative soliton resonance square pulses in all anomalous passively mode-locked fiber laser. *J. Opt.* **2017**, *19*, 105501. [CrossRef]
77. Scott, A. *Nonlinear Science: Emergence and Dynamics*; Oxford University Press: Oxford, UK, 2003.

© 2019 by the authors. Licensee MDPI, Basel, Switzerland. This article is an open access article distributed under the terms and conditions of the Creative Commons Attribution (CC BY) license (http://creativecommons.org/licenses/by/4.0/).

Article

Quantum Behavior of a $\mathcal{PT}$-Symmetric Two-Mode System with Cross-Kerr Nonlinearity

Jan Peřina Jr. [1,*] and Antonín Lukš [2]

1. Joint Laboratory of Optics, Institute of Physics of the Czech Academy of Sciences, 17. listopadu 50a, 771 46 Olomouc, Czech Republic
2. Joint Laboratory of Optics, Faculty of Science, Palacký University, 17. listopadu 12, 771 46 Olomouc, Czech Republic
* Correspondence: jan.perina.jr@upol.cz

Received: 12 July 2019 ; Accepted: 31 July 2019; Published: 7 August 2019

Abstract: Quantum behavior of two oscillator modes, with mutually balanced gain and loss and coupled via linear coupling (including energy conserving as well as energy non-conserving terms) and nonlinear cross-Kerr effect, is investigated. Stationary states are found and their stability analysis is given. Exceptional points for $\mathcal{PT}$-symmetric cases are identified. Quantum dynamics treated by the model of linear operator corrections to a classical solution reveals nonclassical properties of individual modes (squeezing) as well as their entanglement.

Keywords: nonlinearly coupled oscillators; PT symmetry; cross-Kerr nonlinearity; stability analysis; quantum properties

1. Introduction

$\mathcal{PT}$-symmetric systems, which contain gain and loss in mutual balance, have been extensively analyzed in various configurations and from different points of view since the pioneering work by Bender and Boettcher occurred [1–3]. The simplest system is composed of two linearly coupled oscillator modes, one exhibiting gain and the other loss [4]. In real physical applications, there occur additional nonlinear Kerr-type terms in both oscillator modes. They originate in physical models of mode amplification and attenuation typically realized via two-level atoms [5]. These models were developed and extensively discussed in the semiclassical and quantum theories of lasers [6]. Stationary states then occur in such systems due to this nonlinearity. The simplest model of two coupled nonlinear oscillator modes has been generalized to include more oscillators in various configurations. The obtained models were applied in many areas of physics including optical coupled structures [7–10], optical waveguides [11,12], coupled optical micro-resonators [13–15], optical lattices [16–19], opto-mechanical systems [20,21], etc.

Recently, attention has been devoted to the consistent quantum description of $\mathcal{PT}$-symmetric systems. To guarantee the validity of commutation relations among the field operators during the evolution, the fluctuating quantum Langevin forces with specific properties have to be considered in the system [22–25]. As a consequence, the noise in the system constantly increases during the evolution both owing to the amplification and attenuation [24]. Despite this, quantum $\mathcal{PT}$-symmetric systems exhibit interesting and appealing features, such as enhancement of interactions around and at exceptional points (EPs) [26] or quantum Zeno effect [27]. The enhancement of nonlinear interactions then opens the door for the generation of nonclassical light (squeezing) and entangled states [28–31].

Here, we investigate the behavior of a specific form of the model of two coupled oscillator modes with amplification and attenuation that includes only the cross-Kerr nonlinear term. In addition, linear coupling of both modes through $\chi^{(2)}$ parametric interaction that does not conserve energy, is considered [32,33]. It adds or removes photons simultaneously in both modes. This coupling leads to nonclassical properties of the fields and the occurrence of entanglement between the modes [22,32,34], together with the cross-Kerr nonlinear coupling [35,36]. The cross-Kerr coupling is known to play a significant role in quantum non-demolition measurement [37], generation of the states defined in finite-dimensional Hilbert spaces [38] and generation of maximally entangled Bell-type states [39,40]). In general, the cross-Kerr coupling appearing in the so-called Kerr couplers considerably changes their quantum properties [41–43]. The cross-Kerr nonlinearity can even enhance the usual Kerr effect, e.g., when squeezing effects are analyzed [44].

We show that continuous sets of stationary states occur in the model and we analyze their stability. Then, in the framework of the model of quantum superposition of signal and noise [22], we address squeezed-state generation and generation of entangled states around the stationary states. We note that if one of the oscillator modes in the analyzed model attains an additional Kerr nonlinear term, only the trivial stationary states exist. On the other hand, if the standard Kerr nonlinear terms are attributed to both oscillator modes, the system behavior considerably changes and only discrete stationary states are found [45]. We note that related systems were analyzed from the point of view of squeezed-state generation in [29] (without parametric interaction) and [31] (no Kerr terms, parametric interaction in individual modes) and quantum-noise generation [25] (without parametric interaction). In addition, the work [46], where breaking of the oscillatory regime in a classical two-mode $\mathcal{PT}$-symmetric system with the Kerr nonlinearity due to larger modes intensities is reported, is worth mentioning.

The paper is organized as follows. In Section 2, the analyzed system is defined and the corresponding Heisenberg equations are given. Stationary states and their stability are investigated in Section 3. Nonclassical properties of the evolving states are discussed in Section 4. Section 5 presents conclusions.

2. Quantum Hamiltonian and Dynamical Equations

Introducing annihilation ($\hat{a}_j$) and creation ($\hat{a}_j^\dagger$) operators of photons for oscillator modes 1 ($j = 1$) and 2 ($j = 2$) with identical frequencies ω, the considered system is described by the following interaction Hamiltonian $\hat{H}$ [22]:

$$\hat{H} = -i\gamma_1 \hat{a}_1^\dagger \hat{a}_1 - i\gamma_2 \hat{a}_2^\dagger \hat{a}_2 + \left[\epsilon \hat{a}_1^\dagger \hat{a}_2 + \kappa \hat{a}_1 \hat{a}_2 + \text{h.c.}\right] + \beta_c \hat{a}_1^\dagger \hat{a}_2^\dagger \hat{a}_1 \hat{a}_2. \tag{1}$$

We assume that mode 1 is attenuated $\gamma_1 \geq 0$ and mode 2 is amplified $\gamma_2 \leq 0$. Transfer of energy in the system is described by linear coupling constants ϵ and κ. Whereas the coupling constant ϵ characterizes transfer of energy between the modes, the constant κ quantifies energy inserted and removed to/from both modes in the same amount in the $\chi^{(2)}$ parametric process. The nonlinear coupling constant β_c characterizes in Equation (1) the cross-Kerr nonlinear term that is responsible for the occurrence of stationary states. Both $\chi^{(2)}$ term of parametric interaction and cross-Kerr term occur together in nonlinear photonic structures [33] (waveguides, nonlinear fibers). Symbol h.c. replaces the Hermitian conjugated terms.

The Hamiltonian $\hat{H}$ in Equation (1) attains its $\mathcal{PT}$-symmetric form provided that the constants γ_1, γ_2, ϵ, κ and β_c are real and

$$\gamma_1 = -\gamma_2 \equiv \gamma \geq 0. \tag{2}$$

Moreover, to allow for simple physical interpretation, $\mathcal{PT}$-symmetric Hamiltonians are usually applied for the range of parameters in which their linear parts are endowed with real eigenvalues. For the Hamiltonian $\hat{H}$ in Equation (1), this occurs provided that

$$\epsilon^2 - \kappa^2 - \gamma^2 \geq 0. \tag{3}$$

Points in the space of parameters for which equality in Equation (3) holds identify systems with specific properties. They are called exceptional points (EPs) [1]. Without the loss of generality, we further assume $\epsilon > 0$ and $\kappa \geq 0$.

Applying the canonical commutation relations for field operators [5] we obtain the Heisenberg equations from the Hamiltonian $\hat{H}$ in Equation (1),

$$\frac{d\hat{a}_1}{dt} = -\gamma_1\hat{a}_1 - i\epsilon\hat{a}_2 - i\kappa\hat{a}_2^\dagger - i\beta_c\hat{a}_2^\dagger\hat{a}_2\hat{a}_1 + \hat{l}_1,$$

$$\frac{d\hat{a}_2}{dt} = -\gamma_2\hat{a}_2 - i\epsilon\hat{a}_1 - i\kappa\hat{a}_1^\dagger - i\beta_c\hat{a}_1^\dagger\hat{a}_1\hat{a}_2 + \hat{l}_2,$$

(4)

and the Hermitian-conjugated ones. The fluctuating Langevin operator forces $\hat{l}_1$ and $\hat{l}_2$ are introduced in Equation (4) in relation to attenuation in mode 1 and amplification in mode 2, respectively. Their properties [22,23,25],

$$\langle \hat{l}_1^\dagger(t)\hat{l}_1(t')\rangle = 0, \quad \langle \hat{l}_1(t)\hat{l}_1^\dagger(t')\rangle = 2\gamma_1\delta(t-t'),$$
$$\langle \hat{l}_2^\dagger(t)\hat{l}_2(t')\rangle = -2\gamma_2\delta(t-t'), \quad \langle \hat{l}_2(t)\hat{l}_2^\dagger(t')\rangle = 0,$$

(5)

guarantee validity of the commutation relations for the field operators during the evolution. In mode 1, they express the fluctuation-dissipation theorem [47,48] according to which any dissipation of the energy from a system has to be accompanied by back-action from the environment. In analogy, in mode 2, the Langevin forces represent a part of the 'fluctuation-amplification theorem' that occurs as a consequence of consistent adding the energy into the system [6]. Symbol δ means the Dirac function.

3. Stationary States and Their Stability

First, we address the Heisenberg equations in Equation (4) in their 'classical' noiseless limit, i.e., we write them for the coherent states $|\alpha_j\rangle$ with complex amplitudes $\alpha_j = \varrho_j \exp(i\varphi_j)$, $j = 1, 2$:

$$\frac{d\varrho_1}{dt} = -\gamma_1\varrho_1 + [\epsilon \sin(\varphi) - \kappa \sin(\psi)]\varrho_2, \tag{6}$$

$$\frac{d\varrho_2}{dt} = -\gamma_2\varrho_2 - [\epsilon \sin(\varphi) + \kappa \sin(\psi)]\varrho_1, \tag{7}$$

$$\frac{d\varphi_1}{dt} = -[\epsilon \cos(\varphi) + \kappa \cos(\psi)]\frac{\varrho_2}{\varrho_1} - \beta_c\varrho_2^2, \tag{8}$$

$$\frac{d\varphi_2}{dt} = -[\epsilon \cos(\varphi) + \kappa \cos(\psi)]\frac{\varrho_1}{\varrho_2} - \beta_c\varrho_1^2. \tag{9}$$

In Equations (6) and (7), we suitably replace the phases φ_1 and φ_2 by their sum $\psi = \varphi_2 + \varphi_1$ and difference $\varphi = \varphi_2 - \varphi_1$.

To reveal the stationary complex amplitudes α_1 and α_2, we set the time derivatives in Equations (6)–(9) to zero. Before analyzing the obtained equations in detail, we note that there exist the trivial stationary states with $\varrho_1^{st} = \varrho_2^{st} = 0$ and arbitrary values of phases φ_1^{st} and φ_2^{st}. Under the stationary conditions, Equations (8) and (9) for the phases φ_1 and φ_2 are dependent and, e.g., Equation (8) gives us:

$$\varrho_1^{st}\varrho_2^{st} = -(c_\epsilon + c_\kappa)/\beta_c. \tag{10}$$

To simplify the notation, we use in Equation (10) and below the following functions that substitute the phases ψ and φ:

$$s_\epsilon = \epsilon \sin(\varphi^{st}), \quad c_\epsilon = \epsilon \cos(\varphi^{st}), \quad s_\kappa = \kappa \sin(\psi^{st}), \quad c_\kappa = \kappa \cos(\psi^{st}). \tag{11}$$

Furthermore, the coupled Equations (6) and (7) considered to be functions of ϱ_1^{st} and ϱ_2^{st} have a nontrivial solution provided that their determinant is zero:

$$s_\epsilon^2 - s_\kappa^2 + \gamma_1\gamma_2 = 0. \tag{12}$$

Equation (6) when combined with Equation (10) gives us the stationary solution for amplitudes:

$$\varrho^{st}_{1,2} = \sqrt{\frac{\gamma_{2,1}}{\beta_c} \frac{c_\epsilon + c_\kappa}{\pm s_\epsilon + s_\kappa}}. \tag{13}$$

According to Equations (12) and (13), one phase, e.g., ψ^{st} in the stationary solution is free. The other phase, φ^{st}, has to fulfill Equation (12) that admits in general four solutions. The amplitudes $\varrho^{st}_{1,2}$ determined by Equation (13) have to be real and also Equation (10) has to give nonnegative $\varrho^{st}_1 \varrho^{st}_2$.

To address stability of the stationary solution, we derive the linearized equations for deviations $\delta\varrho_1$, $\delta\varrho_2$, $\delta\varphi$ and $\delta\psi$ from their corresponding stationary values ϱ^{st}_1, ϱ^{st}_2, φ^{st}, and ψ^{st}:

$$\frac{d}{dt}\begin{bmatrix}\delta\varrho_1\\ \delta\varrho_2\\ \delta\varphi\\ \delta\psi\end{bmatrix} = \begin{bmatrix}-\gamma_1 & s_\epsilon - s_\kappa & c_\epsilon \varrho^{st}_2 & -c_\kappa \varrho^{st}_2\\ -s_\epsilon - s_\kappa & -\gamma_2 & -c_\epsilon \varrho^{st}_1 & -c_\kappa \varrho^{st}_1\\ G^+ & H^+ & I^+ s_\epsilon & I^+ s_\kappa\\ G^- & H^- & I^- s_\epsilon & I^- s_\kappa\end{bmatrix}\begin{bmatrix}\delta\varrho_1\\ \delta\varrho_2\\ \delta\varphi\\ \delta\psi\end{bmatrix}. \tag{14}$$

The parameters $G^\pm$, $H^\pm$ and $I^\pm$ are given as follows:

$$\begin{aligned}G^\pm &= \frac{\beta_c \varrho^{st}_1}{s_\epsilon - s_\kappa}\left[s_\epsilon\left(\mp\gamma_1/\gamma_2 - 1\right) + s_\kappa\left(\mp\gamma_1/\gamma_2 + 1\right)\right],\\ H^\pm &= \frac{\beta_c \varrho^{st}_2}{s_\epsilon + s_\kappa}\left[s_\epsilon\left(\gamma_2/\gamma_1 \pm 1\right) + s_\kappa\left(-\gamma_2/\gamma_1 \pm 1\right)\right],\\ I^\pm &= (s_\epsilon - s_\kappa)/\gamma_1 \pm (s_\epsilon + s_\kappa)/\gamma_2.\end{aligned} \tag{15}$$

For the $\mathcal{PT}$-symmetric case, eigenvalues ν of the dynamical matrix from Equation (14) are given as roots of the following polynomial:

$$\nu\left(\nu^3 + b\nu + c\right) = 0,$$
$$b = 4(c_\epsilon + c_\kappa)(c_\epsilon s_\kappa^2 + c_\kappa s_\epsilon^2)/\gamma^2, \quad c = -8 s_\epsilon s_\kappa (c_\epsilon + c_\kappa)^2/\gamma. \tag{16}$$

The eigenvalue $\nu_1 = 0$ is related to the freedom in determining, e.g., the phase ψ^{st} of a stationary state. Provided that $c_\epsilon + c_\kappa = 0$, we have $\nu_{1-4} = 0$ and Equation (13) gives us the trivial stationary state $\varrho^{st}_1 = \varrho^{st}_2 = 0$ lying on the border of stability.

Assuming $\mathcal{PT}$-symmetry and special case without $\chi^{(2)}$ interaction ($\kappa = 0$), we have $s_\kappa = c_\kappa = 0$ and the phase ψ^{st} is arbitrary. On the other hand, one solution for the remaining parameters of the stationary state is derived from Equations (12) and (13) for $\beta_c > 0$ in the following implicit form:

$$s_\epsilon = \gamma, \quad c_\epsilon = -\sqrt{\epsilon^2 - \gamma^2}, \quad \varrho^{st}_1 = \varrho^{st}_2 = \sqrt[4]{\epsilon^2 - \gamma^2}/\sqrt{\beta_c}. \tag{17}$$

Similarly, we reveal one stationary solution for $\beta_c < 0$:

$$s_\epsilon = \gamma, \quad c_\epsilon = \sqrt{\epsilon^2 - \gamma^2}, \quad \varrho^{st}_1 = \varrho^{st}_2 = \sqrt[4]{\epsilon^2 - \gamma^2}/\sqrt{-\beta_c}. \tag{18}$$

Eigenvalues of the dynamical matrix in Equation (14) are obtained for both solutions in Equations (17) and (18) as $\nu_{1-4} = 0$, i.e., the states are at the border of stability.

For nonzero κ, we first address the stationary states in EPs ($\mathcal{PT}$-symmetric case) for specific values of the phase ψ^{st}. The trivial stationary states with $\varrho^{st}_1 = \varrho^{st}_2 = 0$ and zero eigenvalues $\nu_{1-4} = 0$ in the stability analysis are found under the conditions summarized in Table 1.

Table 1. Parameters of the trivial stationary states $\varrho_1^{st} = \varrho_2^{st} = 0$ with zero eigenvalues $\nu_{1-4} = 0$ for specific values of the phase ψ^{st}.

$\psi^{st} = 0$	$s_\kappa = 0$	$c_\kappa = \kappa$	$s_\epsilon = \pm\gamma$	$c_\epsilon = -\kappa$
$\psi^{st} = \pi/2$	$s_\kappa = \kappa$	$c_\kappa = 0$	$s_\epsilon = \pm\epsilon$	$c_\epsilon = 0$
$\psi^{st} = \pi$	$s_\kappa = 0$	$c_\kappa = -\kappa$	$s_\epsilon = \pm\gamma$	$c_\epsilon = \kappa$
$\psi^{st} = 3\pi/2$	$s_\kappa = -\kappa$	$c_\kappa = 0$	$s_\epsilon = \pm\epsilon$	$c_\epsilon = 0$

The nontrivial stationary solution with $\varrho_1^{st} = \varrho_2^{st} = \sqrt{2\kappa/\beta_c}$ in EPs is revealed for $\psi^{st} = \pi$ and $\beta_c > 0$:

$$s_\kappa = 0, \quad c_\kappa = -\kappa, \quad s_\epsilon = \gamma, \quad c_\epsilon = -\kappa. \tag{19}$$

On the other hand, we have for $\beta_c < 0$ and $\psi^{st} = 0$ the stationary solution with $\varrho_1^{st} = \varrho_2^{st} = \sqrt{2\kappa/(-\beta_c)}$ in EPs provided that

$$s_\kappa = 0, \quad c_\kappa = \kappa, \quad s_\epsilon = \gamma, \quad c_\epsilon = \kappa. \tag{20}$$

Both solutions in Equations (19) and (20) have the same eigenvalues $\nu_{1,2} = 0$ and $\nu_{3,4} = \pm i2\sqrt{2\kappa}$ in the stability analysis, i.e., no amplification of amplitude fluctuations occur.

For an arbitrary phase ψ^{st}, Equation (12) admits four possible values for the phase φ^{st}:

$$\varphi_1^{st} = \varphi_{base}^{st}, \quad \varphi_2^{st} = \pi - \varphi_{base}^{st}, \quad \varphi_3^{st} = \pi + \varphi_{base}^{st}, \quad \varphi_4^{st} = 2\pi - \varphi_{base}^{st}, \tag{21}$$

where $\varphi_{base}^{st} = \arcsin(\sqrt{\kappa^2\sin^2(\psi^{st}) + \gamma^2}/\epsilon)$. However, only some of them lead to real and nonnegative amplitudes ϱ_1^{st} and ϱ_2^{st} in Equation (13) and nonnegative expression in Equation (10). According to Equation (12), $|s_\epsilon| \geq |s_\kappa|$. Equation (12) can also be recast into the form suitable for the discussion:

$$c_\epsilon^2 - c_\kappa^2 = \epsilon^2 - \kappa^2 - \gamma^2. \tag{22}$$

Considering Equation (22) for $\epsilon^2 - \kappa^2 - \gamma^2 \geq 0$ we have $|c_\epsilon| \geq |c_\kappa|$. If $\beta_c > 0$ [$\beta_c < 0$], Equation (10) requires $c_\epsilon \leq 0$ [$c_\epsilon \geq 0$] and this admits the phases φ_2^{st} and φ_3^{st} [φ_1^{st} and φ_4^{st}]. The expression (13) for ϱ_1^{st} requires $s_\epsilon \geq 0$ independently of the sign of β_c and so only the phases φ_1^{st} and φ_2^{st} can be considered. Both conditions are fulfilled only for the phase φ_2^{st} [φ_1^{st}] for $\beta_c > 0$ [$\beta_c < 0$].

On the other hand, we have $|c_\epsilon| \leq |c_\kappa|$ for $\epsilon^2 - \kappa^2 - \gamma^2 \leq 0$. In this case, $c_\kappa \leq 0$ [$c_\kappa \geq 0$] is needed in Equation (10) for $\beta_c > 0$ [$\beta_c < 0$] and so $\psi^{st} \in \langle \pi/2, 3\pi/2 \rangle$ [$\psi^{st} \in \langle 0, \pi/2 \rangle \cup \langle 3\pi/2, 2\pi \rangle$]. Similarly as above, $s_\epsilon \geq 0$ guarantees nonnegative expression (13) for ϱ_1^{st} for arbitrary β_c, i.e., only the phases φ_1^{st} and φ_2^{st} are allowed. According to both conditions, nontrivial stationary states are expected for the phases φ_1^{st} and φ_2^{st} in the interval of phase $\psi^{st} \in \langle \pi/2, 3\pi/2 \rangle$ [$\psi^{st} \in \langle 0, \pi/2 \rangle \cup \langle 3\pi/2, 2\pi \rangle$] for $\beta_c > 0$ [$\beta_c < 0$].

The EPs occurring at the border of two above discussed regions need special attention. According to Equation (22), we have $|c_\epsilon| = |c_\kappa|$ at the EPs. The analysis reveals that on the top of the stationary states characterized in the above two paragraphs, the trivial stationary states with $\varrho_1^{st} = \varrho_2^{st} = 0$ exist for the phase φ_3^{st} in the interval $\psi^{st} \in \langle 0, \pi/2 \rangle \cup \langle 3\pi/2, 2\pi \rangle$ and for the phase φ_4^{st} in the interval $\psi^{st} \in \langle \pi/2, 3\pi/2 \rangle$ independently of the sign of β_c.

The above general conclusions are further illustrated in the graphs in Figure 1 where the stationary states and their stability are analyzed in the plane $(\gamma/\epsilon, \psi^{st})$ for the case $\beta_c > 0$. In Figure 1, we characterize the stationary states by intensities ϱ_1^{st2} and ϱ_2^{st2} of modes 1 and 2, respectively. We judge the stationary states according to the maximal values of imaginary and real parts of the complex frequencies ν_{1-4}. A positive (negative) imaginary part means amplification (attenuation) of amplitude fluctuations around the stationary state (we note the inverse notation for signs for ν and γ). A nonzero real part then indicates oscillations in the evolution of amplitude fluctuations. We have $\kappa/\epsilon = 0.5$ for the graphs drawn in Figure 1 and so the EP occurs for $\gamma_{EP}/\epsilon = \sqrt{3}/2 \approx 0.87$. The stationary solutions for the phase φ_1^{st} exist only in the area with exponential increase of amplitudes ($\gamma \geq \gamma_{EP}$), they are unstable and amplitude fluctuations oscillate. Only at the EP, the amplitudes ϱ_1^{st} and ϱ_2^{st} are zero and the state is at the border of stability. On the other hand, there exist stationary states for the phase φ_2^{st} in the oscillatory regime of amplitude evolution ($\gamma < \gamma_{EP}$). According to the graph in Figure 1b amplitude fluctuations around these stationary states oscillate. They are amplified except for the line $\psi^{st} = \pi$ that lies at the border of stability (see Figure 1d). According to the graphs in Figure 1e–h, the pattern of intensity ϱ_1^{st2} of mode 1 is a mirror image of the pattern of intensity ϱ_2^{st2} of mode 2 with respect to the plane $\psi^{st} = \pi$ (compare Equation (13) for $\gamma_2 = -\gamma_1$ and $\pm s_\kappa$).

The analysis of the graphs for the case with $\beta_c < 0$ drawn under the conditions of the graphs in Figure 1 reveals similarity provided that we replace φ_1^{st} by φ_2^{st}, shift the phase ψ^{st} by π and exchange mode amplitudes ϱ_1^{st} and ϱ_2^{st}. This similarity is illustrated in the graphs in Figure 2 where the stationary intensities and stability parameters are drawn at the EP for both cases. Considering $\beta_c > 0$ [$\beta_c < 0$] and following the graphs in Figure 2, there exist only the trivial stationary states with $\varrho_1^{st} = \varrho_2^{st} = 0$ at the border of stability for $\psi^{st} \in \langle 0, \pi/2 \rangle \cup \langle 3\pi/2, 2\pi \rangle$ [$\psi^{st} \in \langle \pi/2, 3\pi/2 \rangle$] and φ_2^{st} [φ_1^{st}]. For the remaining phases ψ^{st} nontrivial stationary states are found. These states are unstable except for the state with $\psi^{st} = \pi$ [$\psi^{st} = 0$] and φ_2^{st} [φ_1^{st}] that is at the border of stability and amplitude fluctuations around this state oscillate. Parameters for this state are given in Equation (19) (Equation (20)).

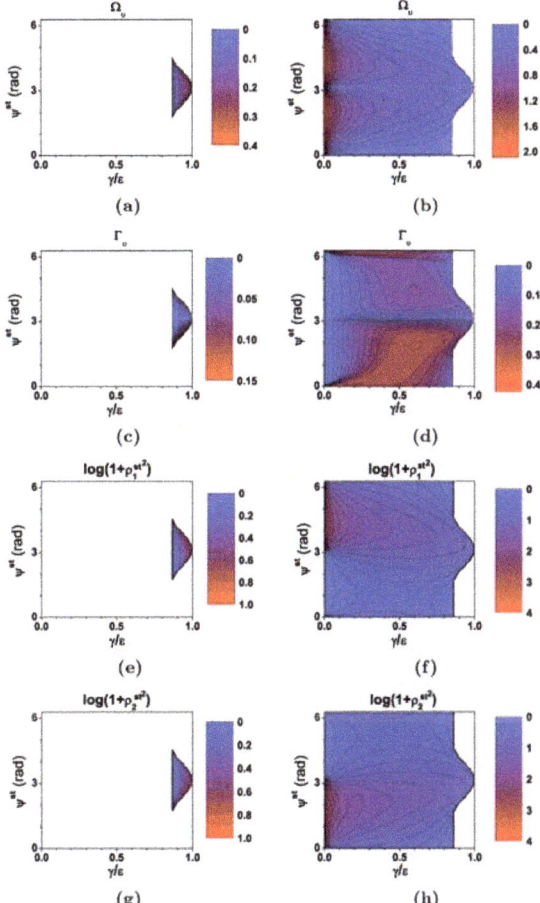

Figure 1. Maximal values of the real and imaginary parts of four complex frequencies ν in the stability analysis expressed via functions $\Omega_\nu = \log[1 + |\text{Re}(\nu)|]$ (**a,b**) and $\Gamma_\nu = \text{sign}[\text{Im}(\nu)] \log[1 + |\text{Im}(\nu)|]$ (**c,d**), respectively, and intensities ϱ_1^{st2} (**e,f**) and ϱ_2^{st2} (**g,h**) of modes 1 and 2, respectively, as they depend on dimensionless attenuation/amplification parameter γ/ϵ and phase ψ^{st} for stationary states with φ_1^{st} (**a,c,e,g**) and φ_2^{st} (**b,d,f,h**) defined in Equation (21); symbol sign gives the sign of the argument, log means the decimal logarithm, Re (Im) stands for the real (imaginary) part of the argument, $\kappa/\epsilon = 0.5$, and $\beta_c/\epsilon = 0.1$.

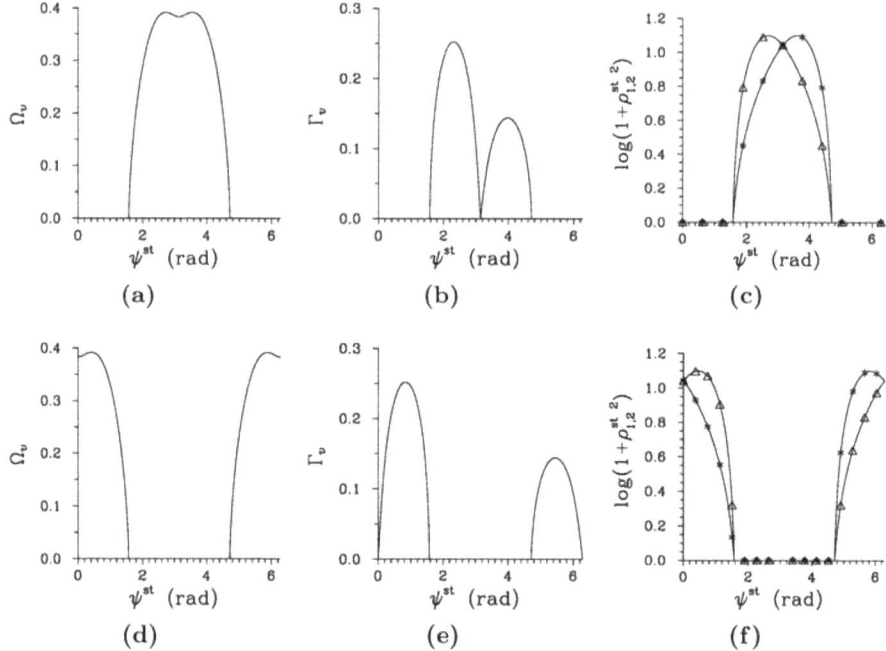

Figure 2. Maximal values of the real and imaginary parts of four complex frequencies ν in the stability analysis expressed by functions Ω_ν (**a,d**) and Γ_ν (**b,e**), respectively, and intensities ϱ^{st2} of mode 1 (*) and 2 ($\triangle$) (**c,f**) as they depend on phase ψ^{st} for φ_2^{st} (see Equation (21)) and $\beta_c/\epsilon = 0.1$ (**a–c**) and φ_1^{st} and $\beta_c/\epsilon = -0.1$ (**d–f**); EP condition $\gamma = \sqrt{\epsilon^2 - \kappa^2}$ is assumed. Functions Ω_ν and Γ_ν as well as the other parameters are given in the caption to Figure 1.

4. Quantum Properties of the Evolving States

We analyze the properties of states evolving from the stationary states determined above and compare them with those characterizing the states originating from non-stationary states. For an initial stationary state, the evolution is described by the Heisenberg equations in Equation (4) linearized around the initial complex amplitudes α_1^{st} and α_2^{st} ($\hat{a}_j = \alpha_j^{st} + \delta\hat{a}_j$, $j = 1, 2$),

$$\frac{d\delta\hat{a}_1}{dt} = -\left(\gamma_1 + i\beta_c|\alpha_2^{st}|^2\right)\delta\hat{a}_1 - i\left(\epsilon + \beta_c\alpha_1^{st}\alpha_2^{st*}\right)\delta\hat{a}_2 - i\left(\kappa + \beta_c\alpha_1^{st}\alpha_2^{st}\right)\delta\hat{a}_2^\dagger + \hat{l}_1,$$

$$\frac{d\delta\hat{a}_2}{dt} = -i\left(\epsilon + \beta_c\alpha_1^{st*}\alpha_2^{st}\right)\delta\hat{a}_1 - i\left(\kappa + \beta_c\alpha_1^{st}\alpha_2^{st}\right)\delta\hat{a}_1^\dagger - \left(\gamma_2 + i\beta_c|\alpha_1^{st}|^2\right)\delta\hat{a}_2 + \hat{l}_2, \qquad (23)$$

and the Hermitian-conjugated ones. When an initial non-stationary state is assumed, we numerically solve the classical nonlinear Equations (6)–(9) and linearize the Heisenberg equations around the evolving complex amplitudes $\alpha_1(t)$ and $\alpha_2(t)$ [45]. In both cases the solution can be expressed in the following general form

$$\begin{bmatrix} \delta\hat{a}_1(t) \\ \delta\hat{a}_2(t) \end{bmatrix} = \mathbf{U}(t)\begin{bmatrix} \delta\hat{a}_1(0) \\ \delta\hat{a}_2(0) \end{bmatrix} + \mathbf{V}(t)\begin{bmatrix} \delta\hat{a}_1^\dagger(0) \\ \delta\hat{a}_2^\dagger(0) \end{bmatrix} + \begin{bmatrix} \hat{f}_1(t) \\ \hat{f}_2(t) \end{bmatrix},$$

$$\begin{bmatrix} \hat{f}_1(t) \\ \hat{f}_2(t) \end{bmatrix} = \int_0^t dt'\, \mathbf{U}(t-t')\begin{bmatrix} \hat{l}_1(t') \\ \hat{l}_2(t') \end{bmatrix} + \int_0^t dt'\, \mathbf{V}(t-t')\begin{bmatrix} \hat{l}_1^\dagger(t') \\ \hat{l}_2^\dagger(t') \end{bmatrix}, \qquad (24)$$

in which the matrices $\mathbf{U}(t)$ and $\mathbf{V}(t)$ and correlation functions of the fluctuating operator forces $\hat{f}_j(t)$, $j = 1, 2$, are determined numerically in general [45] and analytically under specific conditions [35,45].

We consider the initial vacuum state for the evolution of operator amplitude corrections. In this case, the evolving states remain Gaussian and so the following six correlation functions characterize them completely [22,35]:

$$\langle \delta \hat{a}_j^\dagger(t) \delta \hat{a}_j(t) \rangle, \quad \langle [\delta \hat{a}_j(t)]^2 \rangle, \quad j = 1, 2, \quad \langle \delta \hat{a}_1^\dagger(t) \delta \hat{a}_2(t) \rangle, \quad \langle \delta \hat{a}_1(t) \delta \hat{a}_2(t) \rangle. \tag{25}$$

They are easily determined from Equation (24). All quantities characterizing the evolving states can then be expressed in terms of the correlation functions (25). For example, the principal squeeze variance of mode j is obtained as [49]

$$\lambda_j = 1 + 2 \left[\langle \delta \hat{a}_j^\dagger(t) \delta \hat{a}_j(t) \rangle - |\langle [\delta \hat{a}_j(t)]^2 \rangle| \right], \quad j = 1, 2. \tag{26}$$

Determination of the covariance matrix in the symmetric operator ordering then allows to reach the logarithmic negativity E_N that is a suitable quantifier of the entanglement between the modes (for details, see [50,51]).

In Figure 3, we compare the state evolution around a stationary state with that occurring around a non-stationary state. As a stationary state, we consider the state given in Equation (19). The analyzed non-stationary state evolves from the state that differs from that in Equation (19) in the phase $\psi^{\text{init}} = 0$:

$$s_\kappa^{\text{init}} = 0, \quad c_\kappa^{\text{init}} = \kappa, \quad s_\epsilon^{\text{init}} = -\gamma, \quad c_\epsilon^{\text{init}} = \kappa. \tag{27}$$

For the stationary state that lies at the border of stability, both intensities $\varrho_1^2(t)$ and $\varrho_2^2(t)$ increase during the evolution and the fluctuating forces give dominant contribution to this increase (compare solid and dashed curves in Figure 3a). Contrary to this, for the initial non-stationary state the intensity $\varrho_1^2(t)$ of attenuated mode 1 first considerably decreases whereas the intensity $\varrho_2^2(t)$ of amplified mode 2 increases constantly. According to the curves in Figure 3d, the relative contribution of fluctuating forces to the dynamics of intensities is small. In both cases, only the attenuated mode 1 exhibits squeezing (for the principal squeeze variances $\lambda_{1,2}$, see Figure 3b,e) and both modes are entangled (for the logarithmic negativity E_N, see Figure 3c,f) for a limited time period. It is worth noting that both squeezing and entanglement are stronger for the initial non-stationary state. The comparison of curves in Figure 3b,d for the principal squeeze variances $\lambda_{1,2}$ and in Figure 3c,e for the logarithmic negativity E_N drawn with and without the inclusion of fluctuating forces clearly documents substantial role of these forces in consistent description of $\mathcal{PT}$-symmetric quantum systems.

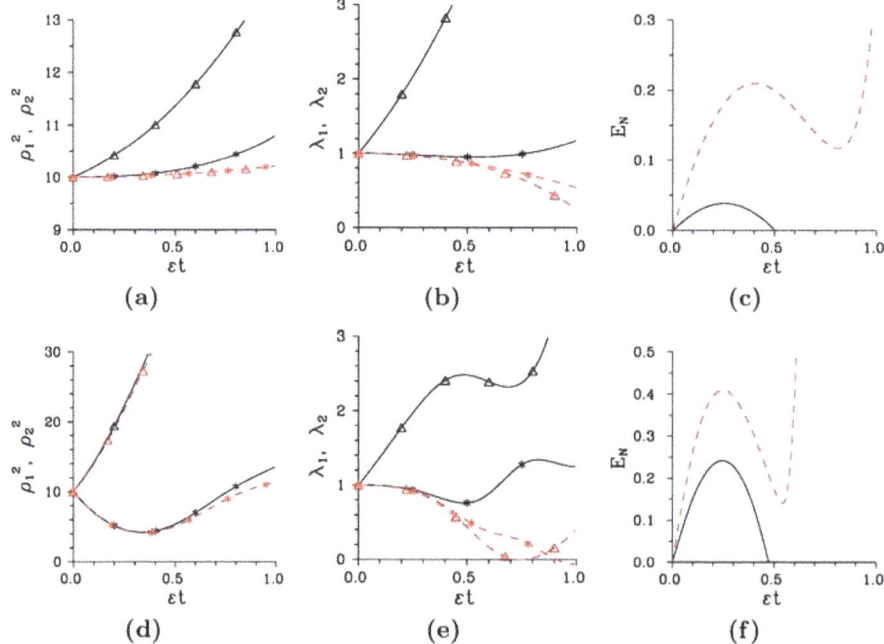

Figure 3. Intensities ϱ^2 (**a,d**) and principal squeeze variances λ of mode 1 ($*$) and 2 ($\triangle$) (**b,e**) and logarithmic negativity E_N (**c,f**) as they evolve along dimensionless time ϵt for initial stationary (non-stationary) state with $\psi^{\rm st} = \pi$ [$\psi^{\rm init} = 0$], $\varrho_{1,2} = \sqrt{2\kappa/\beta_c}$ and $\varphi_2^{\rm st}$ (**a–c**) [$\varphi_4^{\rm init}$ (**c–e**)] given in Equation (19) (Equation (27)), assuming $\gamma = \sqrt{\epsilon^2 - \kappa^2}$, $\kappa/\epsilon = 0.5$, and $\beta_c/\epsilon = 0.1$. The initial vacuum state is assumed, evolution is treated with [without] fluctuating forces (black solid [red dashed] curves).

5. Conclusions

Two oscillator modes with balanced attenuation and amplification were considered to be mutually coupled via the usual linear coupling, $\chi^{(2)}$ parametric process and cross-Kerr nonlinearity. Nontrivial stationary states that occur owing to the cross-Kerr nonlinearity were identified and their stability was determined. The stationary states typically form one-parameter systems. There occur only unstable stationary states and states lying at the border of stability (zero imaginary parts of frequencies in the stability analysis). The solution of linearized operator equations for mode amplitudes around these stationary states revealed nonclassical properties of the evolving states (single-mode squeezing, entanglement). Initial non-stationary states seem to be more suitable for nonclassical-state generation than the stationary ones at the border of stability. Substantial role of the fluctuating Langevin operator forces in consistent description of the system was demonstrated.

Author Contributions: Conceptualization, J.P.J. and A.L.; Investigation, J.P.J. and A.L.; Software, J.P.J.; Writing—original draft, J.P.J.; Writing—review & editing, J.P.J. and A.L.

Funding: J.P. acknowledges the support by the GA ČR project 18-22102S. A.L. gratefully acknowledges the support from the project IGA_PrF_2019_008 of Palacký University.

Acknowledgments: The authors thank J. K. Kalaga and W. Leoński for discussions about PT-symmetry and Kerr effect.

Conflicts of Interest: The authors declare no conflict of interest.

References

1. Bender, C.M.; Boettcher, S. Real Spectra in non-Hermitian Hamiltonians Having $\mathcal{PT}$ Symmetry. *Phys. Rev. Lett.* **1998**, *80*, 5243–5246. [CrossRef]
2. Bender, C.M.; Boettcher, S.; Meisinger, P.N. $\mathcal{PT}$-symmetric quantum mechanics. *J. Math. Phys.* **1999**, *40*, 2201–2229. [CrossRef]
3. Bender, C.M.; Brody, D.C.; Jones, H.F. Must a Hamiltonian be Hermitian? *Am. J. Phys.* **2003**, *71*, 1095–1102. [CrossRef]
4. Morales, J.D.H.; Guerrero, J.; López-Aguayo, S.; Rodríguez-Lara, B.M. Revisiting the Optical $\mathcal{PT}$-symmetric Dimer. *Symmetry* **2016**, *8*, 83. [CrossRef]
5. Meystre, P.; Sargent, M., III. *Elements of Quantum Optics*, 4th ed.; Springer: Berlin, Germany, 2007.
6. Sargent, M.; Scully, M.O.; Lamb, W.E. *Laser Physics*; Addison-Wesley: Boston, MA, USA, 1974.
7. El-Ganainy, R.; Makris, K.G.; Christodoulides, D.N.; Musslimani, Z.H. Theory of coupled optical $\mathcal{PT}$-symmetric structures. *Opt. Lett.* **2007**, *32*, 2632–2634. [CrossRef] [PubMed]
8. Ramezani, H.; Kottos, T.; El-Ganainy, R.; Christodoulides, D.N. Unidirectional nonlinear $\mathcal{PT}$-symmetric optical structures. *Phys. Rev. A* **2010**, *82*, 043803. [CrossRef]
9. Zyablovsky, A.A.; Vinogradov, A.P.; Pukhov, A.A.; Dorofeenko, A.V.; Lisyansky, A.A. $\mathcal{PT}$-symmetry in optics. *Phys.-Uspekhi* **2014**, *57*, 1063–1082. [CrossRef]
10. Ögren, M.; Abdullaev, F.K.; Konotop, V.V. Solitons in a $\mathcal{PT}$-symmetric $\chi^{(2)}$ coupler. *Opt. Lett.* **2017**, *42*, 4079–4082. [CrossRef]
11. Turitsyna, E.G.; Shadrivov, I.V.; Kivshar, Y.S. Guided modes in non-Hermitian optical waveguides. *Phys. Rev. A* **2017**, *96*, 033824. [CrossRef]
12. Xu, X.; Shi, L.; Ren, L.; Zhang, X. Optical gradient forces in $\mathcal{PT}$-symmetric coupled-waveguide structures. *Opt. Express* **2018**, *26*, 10220–10229. [CrossRef]
13. Liu, Z.P.; Zhang, J.; Özdemir, S.K.; Peng, B.; Jing, H.; Lü, X.Y.; Li, C.W.; Yang, L.; Nori, F.; Liu, Y.X. Metrology with $\mathcal{PT}$-Symmetric Cavities: Enhanced Sensitivity near the $\mathcal{PT}$-Phase Transition. *Phys. Rev. Lett.* **2016**, *117*, 110802. [CrossRef] [PubMed]
14. Zhou, X.; Chong, Y.D. $\mathcal{PT}$ symmetry breaking and nonlinear optical isolation in coupled microcavities. *Opt. Express* **2016**, *24*, 6916–6930. [CrossRef] [PubMed]
15. Arkhipov, I.I.; Miranowicz, A.; Di Stefano, O.; Stassi, R.; Savasta, S.; Nori, F.; Özdemir, S.K. Scully-Lamb quantum laser model for parity-time-symmetric whispering-gallery microcavities: Gain saturation effects and nonreciprocity. *Phys. Rev. A* **2019**, *99*, 053806. [CrossRef]
16. Graefe, E.M.; Jones, H.F. $\mathcal{PT}$-symmetric sinusoidal optical lattices at the symmetry-breaking threshold. *Phys. Rev. A* **2011**, *84*, 013818. [CrossRef]
17. Miri, M.A.; Regensburger, A.; Peschel, U.; Christodoulides, D.N. Optical mesh lattices with $\mathcal{PT}$ symmetry. *Phys. Rev. A* **2012**, *86*, 023807. [CrossRef]
18. Ornigotti, M.; Szameit, A. Quasi $\mathcal{PT}$-symmetry in passive photonic lattices. *J. Opt.* **2014**, *16*, 065501. [CrossRef]
19. Shui, T.; Yang, W.X.; Li, L.; Wang, X. Lop-sided Raman-Nath diffraction in $\mathcal{PT}$-antisymmetric atomic lattices. *Opt. Lett.* **2019**, *44*, 2089–2092. [CrossRef]
20. Tchodimou, C.; Djorwe, P.; Nana Engo, S.G. Distant entanglement enhanced in $\mathcal{PT}$-symmetric optomechanics. *Phys. Rev. A* **2017**, *96*, 033856. [CrossRef]
21. Wang, D.Y.; Bai, C.H.; Liu, S.; Zhang, S.; Wang, H.F. Distinguishing photon blockade in a $\mathcal{PT}$-symmetric optomechanical system. *Phys. Rev. A* **2019**, *99*, 043818. [CrossRef]
22. Peřina, J. *Quantum Statistics of Linear and Nonlinear Optical Phenomena*; Kluwer: Dordrecht, The Netherlands, 1991.
23. Agarwal, G.S.; Qu, K. Spontaneous generation of photons in transmission of quantum fields in $\mathcal{PT}$-symmetric optical systems. *Phys. Rev. A* **2012**, *85*, 31802(R). [CrossRef]
24. Scheel, S.; Szameit, A. $\mathcal{PT}$-symmetric photonic quantum systems with gain and loss do not exist. *Eur. Phys. Lett.* **2018**, *122*, 34001. [CrossRef]
25. Peřinová, V.; Lukš, A.; Křepelka, J. Quantum description of a $\mathcal{PT}$-symmetric nonlinear directional coupler. *J. Opt. Soc. Am. B* **2019**, *36*, 855–861. [CrossRef]
26. Antonosyan, D.A.; Solntsev, A.S.; Sukhorukov, A.A. Photon-pair generation in a quadratically nonlinear parity-time symmetric coupler. *Phot. Res.* **2018**, *6*, A6–A9. [CrossRef]

27. Naikoo, J.; Thapliyal, K.; Banerjee, S.; Pathak, A. Quantum Zeno effect and nonclassicality in a $\mathcal{PT}$-symmetric system of coupled cavities. *Phys. Rev. A* **2019**, *99*, 023820. [CrossRef]
28. Miranowicz, A.; Leoński, W. Two-mode optical state truncation and generation of maximally entangled states in pumped nonlinear couplers. *J. Phys. B At. Mol. Opt. Phys.* **2006**, *39*, 1683–1700. [CrossRef]
29. He, B.; Yan, S.B.; Wang, J.; Xiao, M. Quantum noise effects with Kerr-nonlinearity enhancement in coupled gain-loss waveguides. *Phys. Rev. A* **2015**, *91*, 053832. [CrossRef]
30. Kalaga, J.K.; Kowalewska-Kudłaszyk, A.; Leoński, W.; Barasinski, A. Quantum correlations and entanglement in a model comprised of a short chain of nonlinear oscillators. *Phys. Rev. A* **2016**, *94*, 032304. [CrossRef]
31. Vashahri-Ghamsari, S.; He, B.; Xiao, M. Continuous-variable entanglement generation using a hybrid $\mathcal{PT}$-symmetric system. *Phys. Rev. A* **2017**, *96*, 033806. [CrossRef]
32. Mandel, L.; Wolf, E. *Optical Coherence and Quantum Optics*; Cambridge University Press: Cambridge, UK, 1995.
33. Boyd, R.W. *Nonlinear Optics*, 2nd ed.; Academic Press: New York, NY, USA, 2003.
34. Peřina, J., Jr. Coherent light in intense spatiospectral twin beams. *Phys. Rev. A* **2016**, *93*, 063857. [CrossRef]
35. Peřina J., Jr.; Peřina, J. Quantum statistics of nonlinear optical couplers. In *Progress in Optics*; Wolf, E., Ed.; Elsevier: Amsterdam, The Netherlands, 2000; Volume 41, pp. 361–419.
36. Thapliyal, K.; Pathak, A.; Sen, B.; Peřina, J. Higher-order nonclassicalities in a codirectional nonlinear optical coupler: Quantum entanglement, squeezing, and antibunching. *Phys. Rev. A* **2014**, *90*, 013808. [CrossRef]
37. Sanders, B.C.; Milburn, G.J. Complementarity in a quantum nondemolition measurement. *Phys. Rev. A* **1989**, *39*, 694–702. [CrossRef] [PubMed]
38. Leoński, W.; Kowalewska-Kudlaszyk, A. Quantum Scissors and Finite-Dimensional States Engineering. In *Progress in Optics*; Wolf, E., Ed.; Elsevier: Amsterdam, The Netherlands, 2011; Volume 56, pp. 131–185.
39. Kalaga, J.K.; Jarosik, M.W.; Szczęśniak, R.; Nguen, T.D.; Leoński, W. Pulsed Nonlinear Coupler as an Effective Tool for the Bell-Like States Generation. *Acta Phys. Polon. A* **2019**, *135*, 273–275. [CrossRef]
40. Kalaga, J.K.; Kowalewska-Kudłaszyk, A.; Jarosik, M.W.; Szczęśniak, R.; Leoński, W. Enhancement of the entanglement generation via randomly perturbed series of external pulses in a nonlinear Bose-Hubbard dimer. *Nonlinear Dyn.* **2019**. [CrossRef]
41. Korolkova, N.; Peřina, J. Kerr nonlinear coupler with varying linear coupling coefficient. *J. Mod. Opt.* **1997**, *44*, 1525–1534. [CrossRef]
42. Fiurášek, J.; Křepelka, J.; Peřina, J. Quantum-phase properties of the Kerr couplers. *Opt. Commun.* **1999**, *167*, 115–124. [CrossRef]
43. Ariunbold, G.; Peřina, J. Non-classical behaviour and switching in Kerr nonlinear couplers. *J. Mod. Opt.* **2001**, *48*, 1005–1019. [CrossRef]
44. Bartkowiak, M.; Wu, L.A.; Miranowicz, A. Quantum circuits for amplification of Kerr nonlinearity via quadrature squeezing. *J. Phys. B At. Mol. Opt. Phys.* **2014**, *47*, 145501. [CrossRef]
45. Peřina, J., Jr.; Lukš, A.; Kalaga, J.; Leoński, W.; Miranowicz, A. Nonclassical light in quantum $\mathcal{PT}$-symmetric two-mode systems. To be published.
46. Sukhorukov, A.A.; Xu, Z.; Kivshar, Y.S. Nonlinear suppression of time reversals in $\mathcal{PT}$-symmetric optical couplers. *Phys. Rev. A* **2010**, *83*, 043818. [CrossRef]
47. Callen, H.B.; Welton, T.A. Irreversibility and Generalized Noise. *Phys. Rev.* **1951**, *83*, 34–40. [CrossRef]
48. Kubo, R. The fluctuation-dissipation theorem. *Rep. Prog. Phys.* **1966**, *29*, 255–284. [CrossRef]
49. Lukš, A.; Peřinová, V.; Peřina, J. Principal squeezing of vacuum fluctuations. *Opt. Commun.* **1988**, *67*, 149–151. [CrossRef]
50. Hill, S.; Wootters, W.K. Computable entanglement. *Phys. Rev. Lett.* **1997**, *78*, 5022. [CrossRef]
51. Adesso, G.; Illuminati, F. Entanglement in continuous variable systems: Recent advances and current perspectives. *J. Phys. A Math. Theor.* **2007**, *40*, 7821–7880. [CrossRef]

© 2019 by the authors. Licensee MDPI, Basel, Switzerland. This article is an open access article distributed under the terms and conditions of the Creative Commons Attribution (CC BY) license (http://creativecommons.org/licenses/by/4.0/).

Article

Characteristics of the s–Wave Symmetry Superconducting State in the BaGe$_3$ Compound

Kamila A. Szewczyk [1,*], Ewa A. Drzazga-Szczęśniak [2], Marcin W. Jarosik [2], Klaudia M. Szczęśniak [3] and Sandra M. Binek [1]

1. Division of Theoretical Physics, Institute of Physics, Jan Długosz University in Częstochowa, Ave. Armii Krajowej 13/15, 42-200 Częstochowa, Poland
2. Institute of Physics, Częstochowa University of Technology, Ave. Armii Krajowej 19, 42-200 Częstochowa, Poland
3. Faculty of Chemistry, University of Warsaw, Pasteura 1, 02-093 Warsaw, Poland
* Correspondence: kamila.szewczyk@ajd.czest.pl

Received: 29 June 2019; Accepted: 22 July 2019; Published: 1 August 2019

Abstract: Thermodynamic properties of the s–wave symmetry superconducting phase in three selected structures of the BaGe$_3$ compound ($P6_3/mmc$, $Amm2$, and $I4/mmm$) were discussed in the context of DFT results obtained for the Eliashberg function. This compound may enable the implementation of systems for quantum information processing. Calculations were carried out within the Eliashberg formalism due to the fact that the electron–phonon coupling constant falls within the range $\lambda \in \langle 0.73, 0.86 \rangle$. The value of the Coulomb pseudopotential was assumed to be 0.122, in accordance with the experimental results. The value of the Coulomb pseudopotential was assumed to be 0.122, in accordance with the experimental results. The existence of the superconducting state of three different critical temperature values, namely, 4.0 K, 4.5 K and 5.5 K, depending on the considered structure, was stated. We determined the differences in free energy (ΔF) and specific heat (ΔC) between the normal and the superconducting states, as well as the thermodynamic critical field (H_c) as a function of temperature. A drop in the H_c value to zero at the temperature of 4.0 K was observed for the $P6_3/mmc$ structure, which is in good accordance with the experimental data. Further, the values of the dimensionless thermodynamic parameters of the superconducting state were estimated as: $R_\Delta = 2\Delta(0)/k_B T_c \in \{3.68, 3.8, 3.8\}$, $R_C = \Delta C(T_c)/C^N(T_c) \in \{1.55, 1.71, 1.75\}$, and $R_H = T_c C^N(T_c)/H_c^2(0) \in \{0.168, 0.16, 0.158\}$, which are slightly different from the predictions of the Bardeen–Cooper–Schrieffer theory ($[R_\Delta]_{BCS} = 3.53$, $[R_C]_{BCS} = 1.43$, and $[R_H]_{BCS} = 0.168$). This is caused by the occurrence of small retardation and strong coupling effects.

Keywords: s-wave symmetry Eliashberg formalism; BaGe$_3$ superconductor; thermodynamic properties

PACS: 74.20.Fg, 74.25.Bt, 74.62.Fj

---

1. Introduction

Ba–Ge type compounds are intensively studied with respect to the broad range of their possible technological applications, especially in thermoelectric devices [1,2], which are used for energy production and cooling. They are usually designed to acquire energy from the waste heat sources, e.g., from industrial or chemical processes [3]. It should be emphasized here that superconducting properties of the Ba–Ge type compounds are used for the implementation of systems that perform quantum information processing [4].

Relevant information on the various forms of quantum correlations can be found in [5–8]. Particularly noteworthy is [9], in which the arrangement called the nonlinear quantum scissors was proposed for the first time. It allows calculations to be made on quantum states in a Hilbert space with a finite dimension.

Ba–Ge type compounds are well worth the attention with respect to their interesting low-temperature thermodynamic properties, including the s–wave symmetry superconducting ones [10,11]. It was pointed out, in particular, that the $Ba_{24}Ge_{100}$ system undergoes a transition to the superconducting state at the temperature of 0.24 K [12–14], similarly as the $Ba_{24}Si_{100}$ one. This discovery resulted in a great deal of research work concerning the physical properties of the $Ba_{24}Ge_{100}$ material [15,16], which included studies on the interaction of rattling phonons [2]. It is believed that the highly efficient thermoelectric conversion realised by the unusually low thermal conductivity in these materials is closely related to phonons of such a type. Explanation of this concept, as well as its implementation in thermoelectric devices, is one of the most intriguing research problems in the recent twenty years [2,17]. It is worth noting that the superconducting state in the $Ba_{24}Ge_{100}$ compound is also induced by the rattling phonons. Let us take note of the reader, the superconducting properties in functional materials, the structural parameters, the thermal conductivity, and the magnetic properties which are also examined. In this respect, special attention is given to functional oxides [18–20].

In another case, induction of the s–wave symmetry superconducting phase was experimentally observed in the $BaGe_3$ compound of the $P6_3/mmc$ crystalline structure at the temperature of 4 K [21]. The existence of the superconducting state was also stated in compounds of the $CaGe_3$ and the $SiGe_3$ types, for which the critical temperature value fell within the range from 4 K to 7.4 K. The performed ab initio calculations revealed that the materials, which we are going to consider here, belong to the family of s–wave symmetry superconductors with an electron–phonon pairing mechanism [22,23]. The superconductivity in the $BaGe_3$ compound is induced by the relatively strong electron–phonon coupling (indicated by the relatively large value of λ), which compensates for the low value of the logarithmic phonon frequency $\omega_{\ln}$ (see Table 1). The small value of $\omega_{\ln}$ results from the large atomic masses M of Ba and Ge atoms ($\omega_{\ln} \sim 1/\sqrt{M}$). The shape of the Eliashberg function $\alpha^2 F(\omega)$ and the integrated $\lambda(\omega)$ for the $Amm2$ structure were presented in Reference [24]. Phonon modes from various areas contribute uniformly to the increase in λ, which implies the isotropic electron–phonon coupling. Additionally, the high value of the electron density of states at the Fermi level was confirmed, so that the formation of condensate of Cooper pairs is facilitated. Calculations were carried out by means of the VASP program employing the finite displacement method and by means of the ABINIT program using the linear response method. Using the DFT method, one can also model the stoichiometry of the samples (e.g., the functional [25]) in the context of their relationship with magnetic properties.

However, it should be borne in mind that the oxygen excess and deficit can increase and decrease the oxidation degree of cations. The changing of charge state of cations as the consequence of changing of oxygen content changes such magnetic parameters as total magnetic moment and Curie point, as well as such electrical parameters as activation energy and band gap. Moreover, oxygen vacancies affect exchange interactions. Intensity of exchange interactions decreases with oxygen vacancy concentration increasing. Exchange near the oxygen vacancies is negative according to the Goodenough–Kanamori empirical rules. Oxygen vacancies should lead to the formation of the frustration and weak magnetic state such as spin glass [26].

Pressure simulations were additionally performed for the $BaGe_3$ compound. It was found that both the value of the electron–phonon coupling constant (λ) and the value of critical temperature (T_c) decrease with an increase in pressure [24,27]. Interestingly, there are two possible ways of $BaGe_3$ crystallization under high pressure (not exceeding 15 GPa, however), resulting in two different structures, namely, $Amm2$ and $I4/mmm$ [21,24], both remaining metastable even after lowering the pressure down to normal conditions. Two newly discovered structures, $Amm2$ and $I4/mmm$, exhibit some interesting properties.

The superconducting state can be induced in both, at the critical temperature value equal to 4.5 K and 5.5 K, respectively. The *Amm2* phase is dynamically stable. It consists of clusters built of Ge atoms and triangular prisms formed by Ba atoms and intercalated with Ge atoms [24]. Such a structure has not been observed in any other compound of this group, which distinguishes the *Amm2* phase from the two others. The *I4/mmm* structure is very similar to the previously observed structure of the CaGe$_3$ and the XSi$_3$ compounds, where X = Ca, Y or Lu [23]. The Ge$_2$ dimers form square prisms in this crystalline structure. Theoretical predictions point out that the *I4/mmm* phase is dynamically stable under normal conditions.

The latest experimental research concerning the BaGe$_3$ compound was carried out in 2016 [28]. The system was synthesized under high pressure and at high temperature ($p = 15$ GPa, $T = 1300$ K) and the generation of a new structure, classified as *tI32*, was observed. Metallic type electrical conductivity was stated, and the transition to the superconducting state occurred at the temperature of 6.5 K, which still remains the highest recorded critical temperature of this compound.

In the presented work, we are going to determine all the interesting thermodynamical properties of the *s*–wave symmetry superconducting state induced in the BaGe$_3$ compound for three crystalline structures: *P6$_3$/mmc*, *Amm2*, and *I4/mmm*.

Table 1. Selected values of the characteristic parameters of the *s*–wave symmetry superconducting state for the respective structures of the BaGe$_3$ compound (results obtained using Eliashberg's formalism based on DFT data [24]). Thermodynamic parameters appearing in the table have been defined in the text of the work. Their meaning is carefully discussed in [29].

	BaGe$_3$ Compound		
Structure	*P6$_3$/mmc*	*Amm2*	*I4/mmm*
λ	0.73	0.86	0.86
$\omega_{\ln}$ (meV)	10.548	8.21	10.01
T_c (K)	4.0	4.5	5.5
r	0.033	0.047	0.047
R_Δ	3.68	3.8	3.8
R_C	1.55	1.71	1.75
R_H	0.168	0.16	0.158

2. Theoretical Model

The electron–phonon interaction in the BaGe$_3$ compound is relatively strong, which is confirmed by the relatively high values of the electron–phonon coupling constant: $\lambda = 0.73$ for the *P6$_3$/mmc* structure and $\lambda = 0.86$ for the two other structures, i.e. *Amm2* and *I4/mmm* ones [24] (see Table 1). Therefore we used the *s*–wave symmetry Eliashberg formalism, being a generalization of the BCS mean field theory [30,31], to determine the thermodynamic properties of the superconducting state in the considered systems. It should be remembered that the conventional BCS theory studies the results correctly only in the weak electron–phonon coupling limit ($\lambda < 0.3$).

The *s*–wave symmetry Eliashberg equations on the imaginary axis were solved in the self-consistent way for the whole considered temperature range. In the mixed representation, however, we analysed them only for selected temperature values [29,32,33], for which the induction and the extinction of the superconducting state can be most easily observed.

The Eliashberg equations on the imaginary axis take the form of:

$$\varphi_n = \frac{\pi}{\beta} \sum_{m=-M}^{M} \frac{K(i\omega_n - i\omega_m) - \mu^*\theta(\omega_c - |\omega_m|)}{\sqrt{\omega_m^2 Z_m^2 + \varphi_m^2}} \varphi_m, \qquad (1)$$

and
$$Z_n = 1 + \frac{1}{\omega_n}\frac{\pi}{\beta}\sum_{m=-M}^{M}\frac{K(i\omega_n - i\omega_m)}{\sqrt{\omega_m^2 Z_m^2 + \varphi_m^2}}\omega_m Z_m. \qquad (2)$$

The s-wave symmetry order parameter is defined by the ratio: $\Delta_n = \varphi_n/Z_n$, where $\varphi_n = \varphi(i\omega_n)$ represents the order parameter function and $Z_n = Z(i\omega_n)$ is the wave function renormalization factor. Both functions depend directly on the fermionic Matsubara frequency $\omega_n = \pi k_B T(2n-1)$, where k_B denotes the Boltzmann constant. The pairing kernel is given by the formula: $K(z) = 2\int_0^{+\infty} d\omega \frac{\omega}{\omega^2 - z^2}\alpha^2 F(\omega)$, where $\alpha^2 F(\omega)$ is the Eliashberg function. This function models the electron–phonon interaction. It should be noted that the numerical calculations related to the determination of the Eliashberg function were carried out for the ideal crystal structure. In the case of heterogeneity of the system (e.g., crystallites), one can expect the change in the thermodynamic properties of the superconducting phase, if the value of the electron and phonon density of states change. It is also important to change the matrix elements of the electron–phonon interaction. Analogous effects are also observed by examining the magnetic or electric properties of the crystals [34]. In the case of small impurities, Anderson's theorem for the superconducting state with s-wave symmetry decides that the value of T_c will not change [35].

Note that the electron correlations do not contribute to the pairing potential. This means that the order parameter has only s-wave symmetry. The depairing electron correlations are modeled by the Coulomb pseudopotential (μ^*), which can be defined by the formula [36]: $\mu^* = \mu/[1 + \mu \ln(\omega_e/\omega_{ph})]$, where $\mu = \rho(0)U$, with an accuracy of the first order with respect to the the Coulomb potential (U). The symbols $\rho(0)$, ω_e, and ω_{ph} occurring in the formula denote the electronic density of states at the Fermi level, the characteristic electron frequency, and the characteristic phonon frequency, respectively. In our considerations we assumed the experimentally determined value of $\mu^* = 0.122$, which was found for the $P6_3/mmc$ structure. The symbol θ represents the Heaviside function. The value of the characteristic cutoff frequency in the Eliashberg theory should fall within the range $\omega_c \in \langle 3\Omega_{max}, 10\Omega_{max}\rangle$. We assumed $\omega_c = 3\Omega_{max}$ in our calculations. Note that the choice of ω_c does not change the value of the thermodynamic functions that characterize the superconducting state. Only the value of the characteristic phonon frequency (Ω_c) is changed, which is the fitting parameter. The maximum phonon frequency is equal to $\Omega_{max} = 30$ meV for the three crystalline structures considered here.

The Eliashberg function and the Coulomb pseudopotential are two input parameters for the Eliashberg equations. In the considered system, the Eliashberg functions were substituted by the respective coupling constants: $\lambda = 2\int_0^{+\infty} d\omega \alpha^2 F(\omega)/\omega$. There was also introduced the characteristic phonon frequency (Ω_c), which serves as the parameter fitting the model to the experimental data (to the value of the critical temperature).

The Eliashberg equations were solved numerically. We made use of the finite difference approximation of Newton's method and assumed $M = 1100$, like in our other studies [37–40]. We obtained the stability of solutions of the Eliashberg equations within the temperature range from $T_0 = 0.6$ K to T_c.

The Eliashberg equations defined on the imaginary axis allow to determine most of the thermodynamic properties of the superconducting phase, nevertheless, they do not give full information. For the purpose of finding the physical value of the order parameter, the Eliashberg equations should be solved in the mixed representation ($\varphi_n \to \varphi(\omega)$ and $Z_n \to Z(\omega)$). The s-wave symmetry Eliashberg equations in mixed representation can be obtained by the analytic continuation method [33]. They take the form:

$$\varphi(\omega+i\delta) = \pi k_B T \sum_{m=-M}^{M} [K(\omega - i\omega_m) - \mu^*\theta(\omega_c - |\omega_m|)] \frac{\varphi_m}{\sqrt{\omega_m^2 Z_m^2 + \varphi_m^2}}$$
$$+ i\pi \int_0^{+\infty} d\omega' \alpha^2 F(\omega') \left[[f_{BE}(\omega') + f_{FD}(\omega' - \omega)] \times \frac{\varphi(\omega-\omega'+i\delta)}{\sqrt{(\omega-\omega')^2 Z^2(\omega-\omega'+i\delta) - \varphi^2(\omega-\omega'+i\delta)}} \right.$$
$$\left. + i\pi \int_0^{+\infty} d\omega' \alpha^2 F(\omega') \left[[f_{BE}(\omega') + f_{FD}(\omega' + \omega)] \times \frac{\varphi(\omega+\omega'+i\delta)}{\sqrt{(\omega+\omega')^2 Z^2(\omega+\omega'+i\delta) - \varphi^2(\omega+\omega'+i\delta)}} \right], \quad (3)$$

and

$$Z(\omega+i\delta) = 1 + \frac{i}{\omega}\pi k_B T \sum_{m=-M}^{M} K(\omega - i\omega_m) \frac{\omega_m Z_m}{\sqrt{\omega_m^2 Z_m^2 + \varphi_m^2}}$$
$$+ \frac{i\pi}{\omega} \int_0^{+\infty} d\omega' \alpha^2 F(\omega') \left[[f_{BE}(\omega') + f_{FD}(\omega' - \omega)] \times \frac{(\omega-\omega') Z(\omega-\omega'+i\delta)}{\sqrt{(\omega-\omega')^2 Z^2(\omega-\omega'+i\delta) - \varphi^2(\omega-\omega'+i\delta)}} \right.$$
$$\left. + \frac{i\pi}{\omega} \int_0^{+\infty} d\omega' \alpha^2 F(\omega') \left[[f_{BE}(\omega') + f_{FD}(\omega' + \omega)] \times \frac{(\omega+\omega') Z(\omega+\omega'+i\delta)}{\sqrt{(\omega+\omega')^2 Z^2(\omega+\omega'+i\delta) - \varphi^2(\omega+\omega'+i\delta)}} \right], \quad (4)$$

where the symbols $f_{BE}(\omega)$ and $f_{FD}(\omega)$ stand for the Bose–Einstein and the Fermi–Dirac functions, respectively.

3. Numerical Results

We began the analysis of properties of the superconducting state from the determination of the characteristic phonon frequency Ω_c from the equation: $[\Delta_{n=1}(\Omega_c)]_{T=T_c} = 0$. We obtained the following values: 8.5 meV, 6.7 meV, and 8.3 meV, for the $P6_3/mmc$, the $Amm2$, and the $I4/mmm$ structure, respectively. Full numerical results are presented in Figure 1.

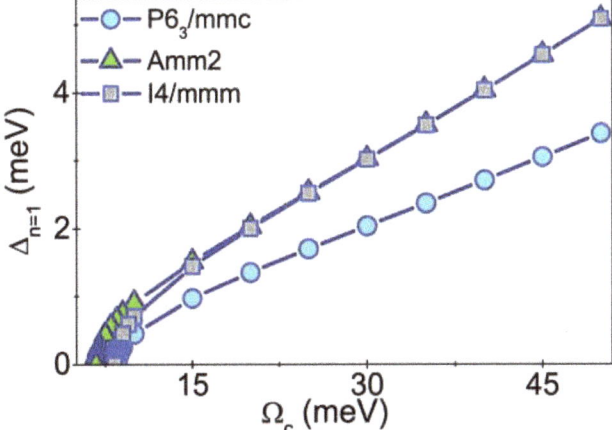

Figure 1. The dependence of the s-wave symmetry order parameter on the characteristic phonon frequency ($T = T_c$).

Then, we solved the s-wave symmetry Eliashberg equations on the imaginary axis. We were interested in the dependence of the order parameter ($\Delta_{n=1}(T)$) and the wave function renormalisation factor ($Z_{n=1}(T)$) on temperature (Figure 2). As it is well known, the order parameter takes zero value for $T \geq T_c$, what allows to estimate the critical temperature. Its respective values were found to be: $T_c = 4.0$ K for $P6_3/mmc$, $T_c = 4.5$ K for $Amm2$, and $T_c = 5.5$ K for $I4/mmm$ structure (see the upper plot in Figure 2). Note that the obtained $\Delta_{n=1}(T)$ functions can be compared with experimental data. As a result, it will be possible to identify the crystal structure of the tested system.

Please note that the results obtained by means of Eliashberg equations can be parameterised in the following way:

$$\Delta_{n=1}(T) = \Delta_{n=1}(0)\sqrt{1 - (T/T_c)^{\gamma}}, \tag{5}$$

where $\gamma = 3.3$ for $P6_3/mmc$ and $\gamma = 3.35$ for both other structures. It is worth mentioning here that the parameter γ is equal to 3.0 in the standard BCS theory [41]. Additionally, $\Delta_{n=1}(0) \in \{0.63, 0.73, 0.89\}$ meV for $P6_3/mmc$, $Amm2$ and $I4/mmm$ structures, respectively.

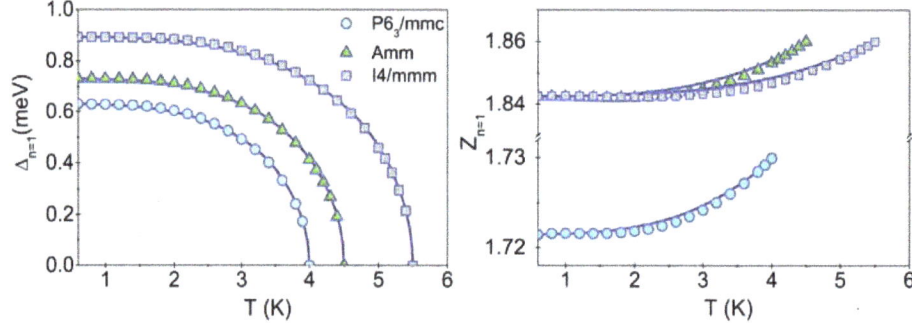

Figure 2. The dependence of the s-wave symmetry order parameter on the temperature (upper plot). The wave function renormalization factor versus temperature (lower plot). Numerical results obtained within the Eliashberg formalism are marked by the point symbols, whereas the solutions obtained on the basis of Equations (5) and (6) are plotted as dark blue solid lines.

The lower plot in Figure 2 presents the wave function renormalization factor, the value of which slightly increases with an increase in temperature and reaches its maximum at T_c. This maximum value should be comparable with the one obtained from the formula: $Z_{n=1}(T_c) = \lambda + 1$. The latter, after obvious calculations, gives the following maximum values for the considered case: $Z_{n=1}(T_c) = 1.73$ for the $P6_3/mmc$ structure and $Z_{n=1}(T_c) = 1.86$ for both other structures under consideration. One can easily see that the maximum values achieved from the numerical solution within the Eliashberg formalism are in good agreement with the ones calculated from the mentioned formula (see Figure 2). Additionally, the full profile of the wave function renormalization factor can be approximately reproduced with the formula:

$$Z_{n=1}(T) = Z_{n=1}(0) + [Z_{n=1}(T_c) - Z_{n=1}(0)]\left(\frac{T}{T_c}\right)^{\gamma}, \tag{6}$$

where $Z_{n=1}(0) = 1.72$ for $P6_3/mmc$ and $Z_{n=1}(0) = 1.84$ for both other structures. Figure 2 indicates the results achieved numerically from the Eliashberg equations by point symbols (triangles, squares, or disks), while the fitting functions are shown as dark blue solid lines.

Subsequently, we determined the temperature dependence of the difference in free energy between the superconducting and the normal state. We used the following formula normalized with respect to the electronic density of states at the Fermi level [42]:

$$\frac{\Delta F}{\rho(0)} = -2\pi k_B T \sum_{n=1}^{M} \left[\sqrt{\omega_n^2 + (\Delta_n)^2} - |\omega_n|\right]$$

$$\times \left[Z_n^{(S)} - Z_n^{(N)} \frac{|\omega_n|}{\sqrt{\omega_n^2 + (\Delta_n)^2}}\right]. \quad (7)$$

Symbols $Z_n^{(S)}$ and $Z_n^{(N)}$ in the above formula represent the wave function renormalization factor for the superconducting (S) and the normal (N) states, respectively.

The negative values of the difference in free energy over the whole temperature range ($T_0 \leq T \leq T_c$), shown in Figure 3 (lower section), confirm the thermodynamic stability of the superconducting state in the crystalline structures under consideration.

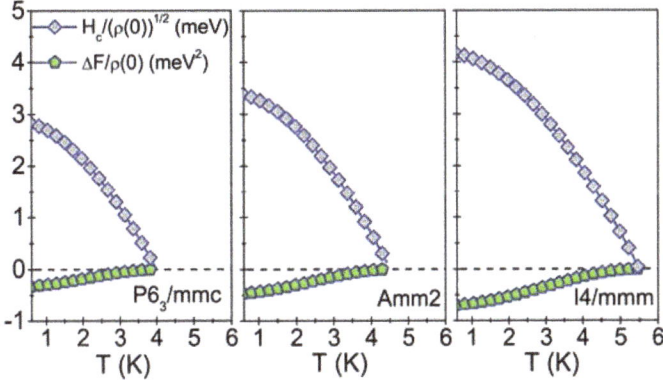

Figure 3. Thermodynamic critical field (upper section) and the difference in free energy between the superconducting and the normal state (lower section) versus temperature.

The acquired knowledge of the temperature dependence of the difference in free energy allowed us to estimate further significant thermodynamic properties of the investigated compound. Our first step was the calculation of the thermodynamic critical field by means of the formula:

$$\frac{H_c}{\sqrt{\rho(0)}} = \sqrt{-8\pi \frac{\Delta F}{\rho(0)}}. \quad (8)$$

Its profile plotted against temperature is shown in Figure 3 (upper section). One can conclude from it that the value of the thermodynamic critical field decreases as the temperature increases, and eventually drops down to zero at T_c. The results presented in Figure 3 are in accordance with the experiment carried out for the $P6_3/mmc$ structure, during which the Meissner effect was observed at the temperature $T = 4.0$ K [21].

Subsequently, the difference in the specific heat ($\Delta C = C^S - C^N$) between the superconducting(S) and the normal (N) state was determined. To do this, we used the formula:

$$\frac{\Delta C(T)}{k_B \rho(0)} = -\frac{1}{\beta} \frac{d^2 [\Delta F/\rho(0)]}{d(k_B T)^2}. \tag{9}$$

The specific heat for the normal state can be determined by means of the formula: $C^N/k_B\rho(0) = \gamma/\beta$, where $\gamma = \frac{2}{3}\pi^2(1+\lambda)$ is the Sommerfeld constant. Numerical results are presented in Figure 4 and indicate linear growth. On the other hand, it can be seen that the specific heat in the superconducting state increases exponentially as the temperature rises. Then, a rapid jump is observed at the critical temperature, followed by further changes proceeding in a way characteristic of the metallic phase. It should be clearly emphasized that the obtained specific heat curves, as well as the thermodynamic critical field curves, can be directly compared with data obtained experimentally.

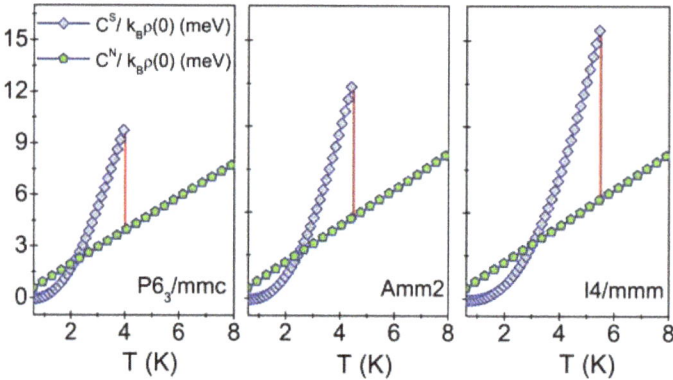

Figure 4. The specific heat for the superconducting state (C^S) and the normal state (C^N) versus temperature. The vertical red lines denote the characteristic jump of the specific heat at T_c.

The succesive stage of calculations consisted in solving the s–wave symmetry Eliashberg equations in mixed representation. In this way, we obtained functions describing both the order parameter and the wave function renormalization factor on the real axis, so that we could analyse the damping effect. The physical value of the order parameter can be obtained by means of the formula: $\Delta(T) = \text{Re}[\Delta(\omega = \Delta(T), T)]$. The results are shown in Figure 5. Each row of charts corresponds to one of the examined structures. The frequency range was assumed to extend from 0 to 50 meV. Such an interval was selected with respect to the construction of the Eliashberg function given for the $Amm2$ structure [24], which takes the non-zero values near (2.25) meV. However, the frequency range for the other structures is remarkably greater, even reaching 45 meV, as can be seen in the charts (Figure 5). The presented plots clearly show that only the real part of the order parameter takes the non-zero values for low frequencies (up to 8 meV). This is evidence of the existence of Cooper pairs with a long lifetime. For higher frequencies, also the imaginary part of the order parameter takes non-zero values. This indicates the existence of Cooper pairs with a finite lifetime. The rapid increase of the damping effects can be seen, especially in the frequency range from about 8 meV to about 30 meV. It is caused by the rather strong electron–phonon coupling within this range.

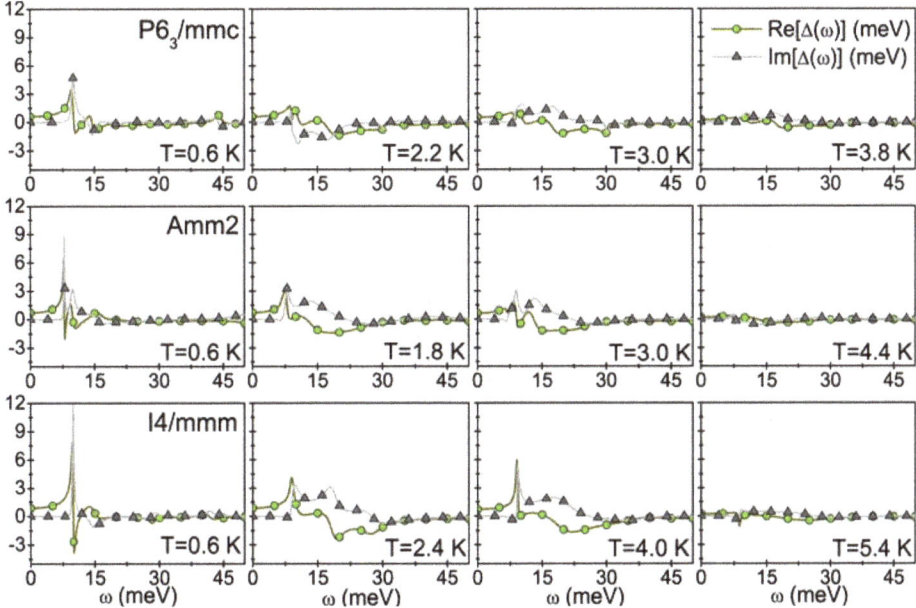

Figure 5. The real and the imaginary part of the s-wave symmetry order parameter for selected temperature values. The first row of charts presents results for the $P6_3/mmc$ structure, the second for $Amm2$, and the third for $I4/mmm$.

The numerical values obtained for the order parameter on the real axis $\text{Re}\left[\Delta(\omega)\right]$ (Figure 5) are in accordance with the values on the imaginary axis $\Delta_{n=1}$ (Figure 2), what confirms the correctness of the performed calculations.

During the last step, we determined the dimensionless termodynamic parameters of the superconducting state: $R_\Delta = 2\Delta(0)/k_B T_c$, $R_C = \Delta C(T_c)/C^N(T_c)$, and $R_H = T_c C^N(T_c)/H_c^2(0)$. One can notice that the results presented in Table 1 for the $P6_3/mmc$ structure are quite close to the predictions of the BCS theory ($[R_\Delta]_{BCS} = 3.53$, $[R_C]_{BCS} = 1.43$, and $[R_H]_{BCS} = 0.168$ [30,31]). Greater divgences can be seen for the two other structures. The apparently different results are directly related to the presence of both the retardation and the strong–coupling effects, which in turn are characterised by the ratio $r = k_B T_c/\omega_{\ln}$. This ratio takes the following values: $r = 0.03$ for the $P6_3/mmc$ structure, and $r = 0.05$ for the two other structures (exact values are given in Table 1). The lack of the above mentioned effects in the BCS theory could be expressed as $r \to 0$.

4. Summary

Slack's predictions about clathrates containing additional atoms as the promising thermoelectric materials [43] encouraged many researchers to seek more such systems. One of them is the $BaGe_3$ compound, for which the induction of the s-wave symmetry superconducting state at the critical temperature of 4.0 K in the $P3_6/mmc$ crystalline structure was observed during experiments [21]. Moreover, two other structures of this compound were discovered on account of investigations, namely, $Amm2$ and $I4/mmm$ [24], synthetized under high pressure and remaining thermodynamically stable under normal conditions, which undergo the s-wave symmetry superconductor–metal phase transition at the temperature of 4.5 K and 5.5 K, respectively. We determined the thermodynamic properties of the superconducting state for the reported structures using the formalism of the s–wave Eliashberg equations by reason of the occurrence of high values of the electron–phonon coupling constants ($\lambda^{P3_6/mmc} = 0.73$, $\lambda^{Amm2} = \lambda^{I4/mmm} = 0.86$). In our calculations, we assumed that $\mu^* = 0.122$, in consistency with the experimental results [21,24]. We determined the thermodynamic functions of the superconducting state which allowed us to find the nondimensional parameters R_Δ, R_C, and R_H. As far as the BCS theory is considered, these parameters are universal constants and their values are as follows: $[R_\Delta]_{BCS} = 3.53$, $[R_C]_{BCS} = 1.43$, and $[R_H]_{BCS} = 0.168$ [30,31]. Within the Eliashberg formalism, we achieved the following results: $R_\Delta = 3.68$, $R_C = 1.55$, and $R_H = 0.168$ for the $P3_6/mmc$ structure; $R_\Delta = 3.8$, $R_C = 1.71$, and $R_H = 0.16$ for the $Amm2$ structure; and finally $R_\Delta = 3.8$, $R_C = 1.75$, and $R_H = 0.158$ for the $I4/mmm$ structure. Our results differ slightly from the predictions of the BCS theory because of the presence of small retardation and strong–coupling effects (the relevant parameter $r = k_B T_c/\omega_{\ln}$ is less than ~ 0.05 for all crystalline structures of the $BaGe_3$ compound).

It should be noted that, in the case of $BaGe_3$, the Eliashberg formalism we used describes the properties of the superconducting state at the quantitative level. This means that the thermodynamic functions are determined very precisely. Due to the relatively low value of the Coulomb pseudopotential, there is no need for any modification of the presented theory.

Author Contributions: Conceptualization, E.A.D.-S.; methodology, M.W.J.; software, E.A.D.-S. and M.W.J.; validation, K.A.S., M.W.J. and E.A.D.-S.; formal analysis, K.A.S.; investigation, K.A.S., K.M.S. and S.M.B.; resources, E.A.D.-S. and M.W.J.; data curation, K.A.S.; visualization, K.M.S. and S.M.B.; supervision, E.A.D.-S.; project administration, E.A.D.-S. and M.W.J.; funding acquisition, K.A.S.

Funding: The results described in the present work were achived with financial support of a Grant for Young Scientist (grant number DSM/WMP/6548/2018) provided by the Jan Długosz University in Częstochowa.

Conflicts of Interest: The authors declare no conflict of interest. The funders had no role in the design of the study; in the collection, analyses, or interpretation of data; in the writing of the manuscript, or in the decision to publish the results.

References

1. Kim, S.J.; Hu, S.; Uher, C.; Hogan, T.; Huang, B.; Corbett, J.D.; Kanatzidis, M.G. Structrure and Thermoelectric Properties of Ba_6Ge_{25-x}, $Ba_6Ge_{23}Sn_2$, and $Ba_6Ge_{22}In_3$: Zintl Phases with a Chiral Clathrate Structure. *J. Solid State Chem.* **2000**, *153*, 321–329. [CrossRef]
2. Paschen, S.; Tran, V.H.; Baenitz, M.; Carrillo-Cabrera, W.; Grin, Y.; Steglich, F. Clathrate Ba_6Ge_{25}: Thermodynamic, magnetic, and transport properties. *Phys. Rev. B* **2002**, *65*, 134435. [CrossRef]
3. Martin, P. Thermoelectric Materials and Applications. *News Bull.* **2005**, *2005*, 30.
4. Gu, X.; Kockum, A.F.; Miranowicz, A.; Liu, Y.X.; Nori, F. Microwave photonics with superconducting quantum circuits. *Phys. Rep.* **2017**, *1*, 718–719. [CrossRef]
5. Kalaga, J.K.; Kowalewska-Kudłaszyk, A.; Leoński, W.; Barasiński, A. Quantum correlations and entanglement in a model comprised of a short chain of nonlinear oscillators. *Phys. Rev. A* **2016**, *94*, 032304. [CrossRef]
6. Kalaga, J.K.; Leoński, W.; Szczęśniak, R. Quantum steering and entanglement in three-mode triangle Bose–Hubbard system. *Quantum Inf. Process.* **2017**, *16*, 265. [CrossRef]
7. Kalaga, J.K.; Leoński, W.; Peřina, J., Jr. Einstein-Podolsky-Rosen steering and coherence in the family of entangled three-qubit states. *Phys. Rev. A* **2018**, *97*, 042110. [CrossRef]
8. Kalaga, J.K.; Leoński, W. Quantum steering borders in three-qubit systems. *Quantum Inf. Process.* **2017**, *16*, 175. [CrossRef]
9. Leoński, W.; Tanaś, R. Possibility of producing the one-photon state in a kicked cavity with a nonlinear Kerr medium. *Phys. Rev. A* **1994**, *49*, R20. [CrossRef]

10. Vaughey, J.T.; Miller, G.J.; Gravelle, S.; Leon-Escamilla, E.A.; Corbett, J.D. Synthesis, Structure, and Properties of $BaGe_2$: A Study of Tetrahedral Cluster Packing and Other Three-Connected Nets in Zintl Phases. *J. Solid State Chem.* **1997**, *133*, 501–507. [CrossRef]
11. Evers, J.; Oehlinger, G.; Ott, H.R. Superconductivity of $SrSi_2$ and $BaGe_2$ with the $\alpha - ThSi_2$-type structure. *J. Less-Common Met.* **1980**, *69*, 389–391. [CrossRef]
12. Fukuoka, H.; Ueno, K.; Yamanaka, S. High-pressure synthesis and structure of a new silicon clathrate $Ba_{24}Si_{100}$. *J. Organomet. Chem.* **2000**, *611*, 543–546. [CrossRef]
13. Grosche, F.M.; Yuan, H.Q.; Carrillo-Cabrera, W.; Paschen, S.; Langhammer, C.; Kromer, F.; Sparn, G.; Baenitz, M.; Grin, Y.; Steglich, F. Superconductivity in the Filled Cage Compounds Ba_6Ge_{25} and $Ba_4Na_2Ge_{25}$. *Phys. Rev. Lett.* **2001**, *87*, 247003. [CrossRef] [PubMed]
14. Yuan, H.Q.; Grosche, F.M.; Carrillo-Cabrera, W.; Pacheco, V.; Sparn, G.; Baenitz, M.; Schwarz, U.; Grin, Y.; Steglich, F. Interplay of superconductivity and structural phase transition in the clathrate Ba_6Ge_{25}. *Phys. Rev. Lett.* **2004**, *70*, 174512.
15. Kanetake, F.; Harada, A.; Mukuda, H.; Kitaoka, Y.; Rachi, T.; Tanigaki, K.; Itoh, K.M.; Haller, E.E. ^{73}Ge- and $^{135/137}$Ba-NMR Studies of Clathrate Superconductor $Ba_{24}Ge_{100}$. *J. Phys. Soc. Jpn.* **2009**, *78*, 104710. [CrossRef]
16. Tang, J.; Xu, J.; Heguri, S.; Fukuoka, H.; Yamanaka, S.; Akai, K.; Tanigaki, K. Electron-Phonon Interactions of Si_{100} and Ge_{100} Superconductors with Ba Atoms Inside. *Phys. Rev. Lett.* **2010**, *105*, 176402. [CrossRef] [PubMed]
17. Shimono, Y.; Shibauchi, T.; Kasahara, Y.; Kato, T.; Hashimoto, K.; Matsuda, Y.; Yamaura, J.; Nagao, Y.; Hiroi, Z. Effects of Rattling Phonons on the Dynamics of Quasiparticle Excitation in the β-Pyrochlore KOs_2O_6 Superconductor. *Phys. Rev. Lett.* **2007**, *98*, 257004. [CrossRef] [PubMed]
18. Trukhanov, A.V.; Kostishyn, V.G.; Panina, L.V.; Korovushkin, V.V.; Turchenko, V.A.; Thakur, P.; Thakur, A.; Yang, Y.; Vinnik, D.A.; Yakovenko, E.S.; et al. Control of electromagnetic properties in substituted M-type hexagonal ferrites. *J. Alloys Compd.* **2018**, *754*, 247–256. [CrossRef]
19. Trukhanov, A.V.; Trukhanov, S.V.; Kostishyn, V.G.; Panina, L.V.; Korovushkin, V.V.; Turchenko, V.A.; Vinnik, D.A.; Yakovenko, E.S.; Zagorodnii, V.V.; Launetz, V.L.; et al. Correlation of the atomic structure, magnetic properties and microwave characteristics in substituted hexagonal ferrites. *J. Magn. Magn. Mater.* **2018**, *462*, 127–135. [CrossRef]
20. Trukhanov, A.V.; Kozlovskiy, A.L.; Ryskulov, A.E.; Uglov, V.V.; Kislitsin, S.B.; Zdorovets, M.V.; Trukhanov, S.V.; Zubar, T.I.; Astapovich, K.A.; et al. Control of structural parameters and thermal conductivity of BeO ceramics using heavy ion irradiation and post-radiation annealing. *Ceram. Int.* **2019**, *45*, 15412–15416. [CrossRef]
21. Fukuoka, H.; Tomomitsu, Y.; Inumaru, K. High-Pressure Synthesis and Superconductivity of a New Binary Barium Germanide $BaGe_3$. *Inorg. Chem.* **2009**, *50*, 6372–6377. [CrossRef] [PubMed]
22. Schnelle, W.; Ormeci, A.; Wosylus, A.; Meier, K.; Grin, Y.; Schwarz, U. Dumbbells of Five-Connected Ge Atoms and Superconductivity in $CaGe_3$. *Inorg. Chem.* **2012**, *51*, 5509–5511. [CrossRef] [PubMed]
23. Schwarz, U.; Wosylus, A.; Rosner, H.; Schnelle, W.; Ormeci, A.; Meier, K.; Baranov, A.; Nicklas, M.; Leipe, S.; Müller, C.J.; et al. Dumbbells of Five-Connected Silicon Atoms and Superconductivity in the Binary Silicides MSi_3 ($M = Ca, Y, Lu$). *J. Am. Chem. Soc.* **2012**, *134*, 13558–13561. [CrossRef] [PubMed]
24. Zurek, E.; Yao, Y. Theoretical Predictions of Novel Superconducting Phases of $BaGe_3$ Stable at Atmospheric and High Pressures. *Inorg. Chem.* **2015**, *54*, 2875. [CrossRef] [PubMed]
25. Trukhanov, S.V.; Trukhanov, A.V.; Vasiliev, A.N.; Szymczak, H. Frustrated Exchange Interactions Formation at Low Temperatures and High Hydrostatic Pressures in $La_{0.70}Sr_{0.30}MnO_{2.85}$. *J. Exp. Theor. Phys.* **2019**, *111*, 209. [CrossRef]
26. Trukhanov, S.V.; Trukhanov, A.V.; Botez, C.E.; Adair, A.H.; Szymczak, H.; Szymczak, R. Magnetic State of the Structural Separated Anion Deficient $La_{0.70}Sr_{0.30}MnO_{2.85}$ Manganite. *J. Exp. Theor. Phys.* **2019**, *113*, 819. [CrossRef]
27. Chen, X.J.; Zhang, C.; Meng, Y.; Zhang, R.Q.; Lin, H.Q.; Struzhkin, V.V.; Mao, H. $\beta - tin \rightarrow Imma \rightarrow sh$ Phase Transitions of Germanium. *Phys. Rev. Lett.* **2011**, *106*, 135502. [CrossRef]
28. Castillo, R.; Baranov, A.I.; Burkhardt, U.; Cardoso-Gil, R.; Schnelle, W.; Bobnar, M.; Schwarz, U. Germanium Dumbbells in a New Superconducting Modification of $BaGe_3$. *Inorg. Chem.* **2016**, *55*, 4498. [CrossRef]

29. Carbotte, J.P. Properties of boson-exchange superconductors. *Rev. Mod. Phys.* **1990**, *62*, 1027. [CrossRef]
30. Bardeen, J.; Cooper, L.N.; Schrieffer, J.R. Microscopic theory of superconductivity. *Phys. Rev.* **1957**, *106*, 162. [CrossRef]
31. Bardeen, J.; Cooper, L.N.; Schrieffer, J.R. Theory of superconductivity. *Phys. Rev.* **1957**, *108*, 1175. [CrossRef]
32. Eliashberg, G.M. Interactions between electrons and lattice vibrations in a superconductor. *Sov. Phys. JETP* **1960**, *11*, 696.
33. Marsiglio, F.; Schossmann, M.; Carbotte, J.P. Iterative analytic continuation of the electron self-energy to the real axis. *Phys. Rev. B* **1988**, *37*, 4965. [CrossRef] [PubMed]
34. Trukhanov, S.V.; Trukhanov, A.V.; Botez, C.E.; Adair, A.H.; Szymczak, H.; Szymczak, R. Phase separation and size effects in $Pr_{0.70}Ba_{0.30}MnO_{3+\delta}$ perovskite manganites. *J. Phys. Condens. Matter* **2007**, *19*, 266214. [CrossRef] [PubMed]
35. Anderson, P.W. Knight Shift in Superconductors. *Phys. Rev. Lett.* **1959**, *3*, 325. [CrossRef]
36. Morel, P.; Anderson, P.W. Calculation of the superconducting state parameters with retarded electron-phonon interaction. *Phys. Rev.* **1962**, *125*, 1263. [CrossRef]
37. Durajski, A.P.; Szczęśniak, R.; Li, Y. Non-BCS thermodynamic properties of H_2S superconductor. *Phys. C Supercond. Its Appl.* **2015**, *515*, 1–6. [CrossRef]
38. Szczęśniak, D.; Szczęśniak, R. Thermodynamics of the hydrogen dominant potassium hydride superconductor at high pressure. *Solid State Commun.* **2015**, *212*, 1. [CrossRef]
39. Duda, A.M.; Szewczyk, K.A.; Jarosik, M.W.; Szczęśniak, K.M.; Sowińska, M.A.; Szczęśniak, D. Characterization of the superconducting state in hafnium hydride under high pressure. *Phys. B Condens. Matter* **2018**, *536*, 275. [CrossRef]
40. Szczęśniak, D.; Kaczmarek, A.Z.; Szczęśniak, R.; Turchuk, S.V.; Zhao, H.; Drzazga, E.A. Superconducting properties of under- and over-doped $Ba_xK_{1-x}BiO_3$ perovskite oxide. *Mod. Phys. Lett. B* **2018**, *32*, 1850174. [CrossRef]
41. Eschrig, H. *Theory of Superconductivity: A Primer*; Citeseer: State College, PA, USA, 2001.
42. Bardeen, J.; Stephen, M. Free-energy difference between normal and superconducting states. *Phys. Rev.* **1964**, *136*, A1485. [CrossRef]
43. Slack, G.A. *CRC Handbook of Thermoelectrics*; Character 34; Rowe, D.M., Ed.; Chemical Rubber: Boca Raton, FL, USA, 1995.

© 2019 by the authors. Licensee MDPI, Basel, Switzerland. This article is an open access article distributed under the terms and conditions of the Creative Commons Attribution (CC BY) license (http://creativecommons.org/licenses/by/4.0/).

Article

The Entanglement Generation in $\mathcal{PT}$-Symmetric Optical Quadrimer System

Joanna K. Kalaga [1,2]

[1] Quantum Optics and Engineering Division, Institute of Physics, University of Zielona Góra, Prof. Z. Szafrana 4a, 65-516 Zielona Góra, Poland; j.kalaga@if.uz.zgora.pl
[2] Joint Laboratory of Optics of Palacký University and Institute of Physics of CAS, Faculty of Science, Palacký University, 17. listopadu 12, 771 46 Olomouc, Czech Republic

Received: 12 July 2019; Accepted: 29 August 2019; Published: 3 September 2019

Abstract: We discuss a model consisting of four single-mode cavities with gain and loss energy in the first and last modes. The cavities are coupled to each other by linear interaction and form a chain. Such a system is described by a non-Hermitian Hamiltonian which, under some conditions, becomes $\mathcal{PT}$-symmetric. We identify the phase-transition point and study the possibility of generation bipartite entanglement (entanglement between all pairs of cavities) in the system.

Keywords: $\mathcal{PT}$-symmetry; entanglement; negativity

1. Introduction

One of the main principles of quantum mechanics is the assumption that the Hamiltonian describing a quantum system must be Hermitian. In consequence, all of Hamiltonian's eigenvalues are real. In 1998, Bender and Boettcher [1] showed that the Hermiticity of a Hamiltonian ($\hat{H} = \hat{H}^\dagger$) is not the only condition to obtain its real eigenvalues. When non-Hermitian Hamiltonian has $\mathcal{PT}$ symmetry, its eigenvalues are real. $\mathcal{PT}$ symmetry means that the Hamiltonian satisfies commutation relations $[\hat{H}, \hat{P}\hat{T}] = 0$, where $\hat{P}$ is a linear parity operator which changes the sign of the momentum operator and the position operator; whereas $\hat{T}$ is the antilinear time reversal operator.

The first of the studied $\mathcal{PT}$-symmetric Hamiltonians with a real spectrum was that described by the equation $\hat{H} = \hat{p}^2 + i\hat{x}^3$ [1–3]. The most commonly discussed quantum mechanical systems with $\mathcal{PT}$ symmetry are those described by Hamiltonians including a complex potential. When the imaginary part has a plus sign, the system obtains energy from the environment; whereas the minus sign means that the system gives energy into the environment. When the system satisfies $\mathcal{PT}$ symmetry, the loss and gain must be balanced. For optical $\mathcal{PT}$-symmetric systems, the refractive index n can play a role of the potential energy V [4]. Quantum optical systems should also exhibit such balance between the gain and loss of energy to satisfy $\mathcal{PT}$ symmetry condition. For such a case, the refractive index obeys the symmetry relation $n(x) = n^*(-x)$.

In recent years, the different kinds of $\mathcal{PT}$-symmetric systems have been considered. For instance, there has been the pair of coupled LC circuits [5], optical lattice [6,7], optomechanical system [8], optical waveguides [9], quantum-dot [10], and others. The $\mathcal{PT}$-symmetric systems can exhibit numerous interesting phenomena such as invisibility [11], chaos induced by the $\mathcal{PT}$ symmetry breaking [12], and many others.

For the $\mathcal{PT}$-symmetric system, we can observe a phase-transition point which is the point where the system loses its $\mathcal{PT}$-symmetric properties. If the system is in the $\mathcal{PT}$-symmetric phase, all eigenvalues of Hamiltonian are real. When the system is in $\mathcal{PT}$ symmetry broken phase, it has complex eigenvalue spectra. Such a transition point from the unbroken $\mathcal{PT}$-symmetric phase to the $\mathcal{PT}$ symmetry broken phase is called exceptional point [13–15]. In general, this singular point occurs when eigenvalues and corresponding eigenvectors of the system depend on some parameters. When those parameters reach a critical value, then eigenvalues of the system coalesce and the spectrum becomes complex. If two eigenvalues coalescence, we have second-order exceptional points. During recent years, various type of singularities and their features have been studied [16–22]. The $\mathcal{PT}$-symmetric systems exhibit many interesting phenomena related to the presence of an exceptional point. For instance, enhancing spontaneous emission [23], unidirectional invisibility in fiber networks [24], and loss-induced lasing [25].

In this paper, we concentrate on the entangled states generation in the $\mathcal{PT}$-symmetric quadrimer system. In particular, we are interested in producing bipartite entangled states, and the influence of gain and loss rate on producing such states. The generation of entanglement in various quantum systems is one of the fundamental areas of interest in quantum information theory. Entanglement can be observed in various physical systems such as Bose–Einstein condensates [26,27], cavity QED [28], quantum dots [29,30], trapped ions [31], and many others [32–37]. The research related to the production of entanglement in open systems is of particular importance. For such a system, a crucial role is played by the decoherence processes, which are the consequence of the interaction of the system with the environment. One should realize that the entanglement is very sensitive to the noise processes. The interaction with the environment leads to losses of coherence, and, in consequence, destroys entanglement. Very interesting is sudden death and the rebirth of entanglement observed in various systems interacting with the environment [38–41].

The paper is organized as follows. In Section 2, we describe the $\mathcal{PT}$-symmetric system. In particular, we derive the formulas determining the location of the exceptional point, the point where the system loses its $\mathcal{PT}$ symmetry, then, the eigenvalues of Hamiltonian become complex. For the different values of gain/loss rate, we analyze the possibility of generation entanglement state in Section 3. We show how the gain/loss rate influences the entanglement production process.

2. The Model

The considered system is composed of four identical single-mode cavities of the resonance frequency ω. The first cavity is passive (it loses the energy), the second and third ones are neutral (no losses and no gain of the energy), and the last one is active (it gains the energy). All cavities are coupled mutually by linear interaction and form a chain (see Figure 1). Such a system can be described by the following Hamiltonian:

$$\begin{aligned}\hat{H} &= (\omega - i\gamma)\,\hat{a}_1^\dagger \hat{a}_1 + \omega \hat{a}_2^\dagger \hat{a}_2 + \omega \hat{a}_3^\dagger \hat{a}_3 + (\omega + i\gamma)\,\hat{a}_4^\dagger \hat{a}_4 \\ &\quad + \beta\left(\hat{a}_1^\dagger \hat{a}_2 + \hat{a}_2^\dagger \hat{a}_1 + \hat{a}_2^\dagger \hat{a}_3 + \hat{a}_3^\dagger \hat{a}_2 + \hat{a}_3^\dagger \hat{a}_4 + \hat{a}_4^\dagger \hat{a}_3\right),\end{aligned} \quad (1)$$

where $\hat{a}_i$ and $\hat{a}_i^\dagger$ are bosonic annihilation and creation operators, respectively, whereas the indices 1, 2, 3, 4 label four subsystems. The parameter β describes the strength of linear interaction between two nearest cavities and γ denotes the strength of decay or the gain of cavities. We assume here that $\hbar = 1$ and the parameters ω, γ, and β are real.

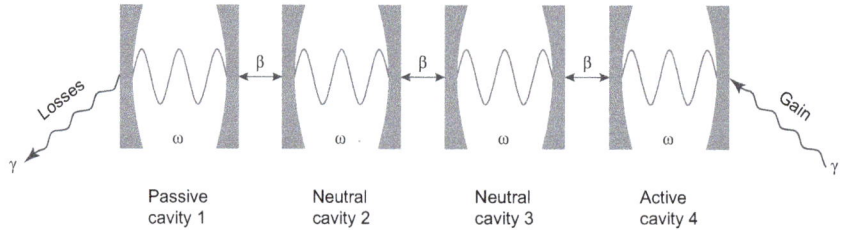

Figure 1. Scheme of the model.

The Hamiltonian defined in Equation (1) is non-Hermitian, but for some situations it becomes $\mathcal{PT}$-symmetric and its eigenvalues are real. To find the $\mathcal{PT}$ phase transition point, we write the Hamiltonian $\hat{H}$ in this form:

$$\hat{H} = \begin{bmatrix} \hat{a}_1^\dagger & \hat{a}_2^\dagger & \hat{a}_3^\dagger & \hat{a}_4^\dagger \end{bmatrix} \mathcal{H} \begin{bmatrix} \hat{a}_1 \\ \hat{a}_2 \\ \hat{a}_3 \\ \hat{a}_4 \end{bmatrix}, \qquad (2)$$

where

$$\mathcal{H} = \begin{bmatrix} \omega - i\gamma & \beta & 0 & 0 \\ \beta & \omega & \beta & 0 \\ 0 & \beta & \omega & \beta \\ 0 & 0 & \beta & \omega + i\gamma \end{bmatrix}. \qquad (3)$$

Next, we can calculate the eigenvalues of the Hamiltonian $\mathcal{H}$, and they are

$$\begin{aligned}
E_1 &= \tfrac{1}{2}\left(2\omega - \sqrt{2}\sqrt{3\beta^2 - \sqrt{5\beta^4 - 2\beta^2\gamma^2 + \gamma^4} - \gamma^2}\right), \\
E_2 &= \tfrac{1}{2}\left(2\omega + \sqrt{2}\sqrt{3\beta^2 - \sqrt{5\beta^4 - 2\beta^2\gamma^2 + \gamma^4} - \gamma^2}\right), \\
E_3 &= \tfrac{1}{2}\left(2\omega - \sqrt{2}\sqrt{3\beta^2 + \sqrt{5\beta^4 - 2\beta^2\gamma^2 + \gamma^4} - \gamma^2}\right), \\
E_4 &= \tfrac{1}{2}\left(2\omega + \sqrt{2}\sqrt{3\beta^2 + \sqrt{5\beta^4 - 2\beta^2\gamma^2 + \gamma^4} - \gamma^2}\right).
\end{aligned} \qquad (4)$$

We see that all eigenvalues depend on the frequency ω, the strength of coupling β, and the gain and loss rate γ. The eigenvalues are real when two relations are satisfied simultaneously: $\left(3\beta^2 - \sqrt{5\beta^4 - 2\beta^2\gamma^2 + \gamma^4} - \gamma^2\right) \geq 0$ and $\left(3\beta^2 + \sqrt{5\beta^4 - 2\beta^2\gamma^2 + \gamma^4} - \gamma^2\right) \geq 0$.

In Figure 2a,b the real and imaginary parts of E_i are presented, respectively. We see there that the phase-transition point is localized when $\gamma = \beta$. If the gain/loss rate γ is smaller than the coupling parameter β, the spectrum is real and the system is in the $\mathcal{PT}$-symmetric phase. As the parameter γ exceeds β, the system passes into the broken $\mathcal{PT}$-symmetric phase, and the eigenvalues become complex. Observed here, the phase-transition point is the second-order exceptional point at which the two eigenvalues coalesce.

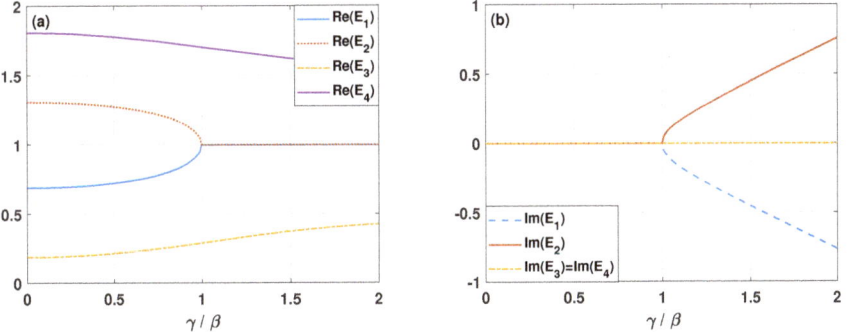

Figure 2. The values of (a) the real part of E_i and (b) the imaginary part of E_i, as a function of ratio γ/β. All parameters are scaled in ω units and $\beta = \omega/2$.

To study the system's dynamics, we analyze the time evolution of the density matrix $\hat{\rho}$. We can note that $\mathcal{PT}$-symmetric system is an open system which exchanges energy with the environment. The energy is gained and lost by the system. The evolution of such a system is determined by a master equation for the operator $\hat{\rho}$

$$\frac{d}{dt}\hat{\rho} = -i\left[\hat{H}_0, \hat{\rho}\right] + \mathcal{L}\hat{\rho}, \tag{5}$$

where

$$\hat{H}_0 = \omega \left(\hat{a}_1^\dagger \hat{a}_1 + \hat{a}_2^\dagger \hat{a}_2 + \hat{a}_3^\dagger \hat{a}_3 + \hat{a}_4^\dagger \hat{a}_4\right) + \beta \left(\hat{a}_1^\dagger \hat{a}_2 + \hat{a}_2^\dagger \hat{a}_1 + \hat{a}_2^\dagger \hat{a}_3 + \hat{a}_3^\dagger \hat{a}_2 + \hat{a}_3^\dagger \hat{a}_4 + \hat{a}_4^\dagger \hat{a}_3\right), \tag{6}$$

and $\mathcal{L}$ is the Liouvillian superoperator, which in our case it takes the form:

$$\mathcal{L}\hat{\rho} = \gamma \left(2\hat{a}_1 \hat{\rho} \hat{a}_1^\dagger - \hat{a}_1^\dagger \hat{a}_1 \hat{\rho} - \hat{\rho} \hat{a}_1^\dagger \hat{a}_1\right) + \gamma \left(2\hat{a}_4^\dagger \hat{\rho} \hat{a}_4 - \hat{a}_4 \hat{a}_4^\dagger \hat{\rho} - \hat{\rho} \hat{a}_4 \hat{a}_4^\dagger\right). \tag{7}$$

In Equation (7), the first term describes the loss and the second term is related to the gain of energy.

In further considerations, we assume that initially only one of the four cavities is in the one-photon state $|1\rangle$, and the other three are in vacuum state $|0\rangle$. Therefore, we will discuss here four cases:

- $\hat{\rho}(t=0) = |1000\rangle\langle 1000|$;
- $\hat{\rho}(t=0) = |0010\rangle\langle 0010|$;
- $\hat{\rho}(t=0) = |0100\rangle\langle 0100|$;
- $\hat{\rho}(t=0) = |0001\rangle\langle 0001|$,

where $|ijkl\rangle = |i\rangle_1 \otimes |j\rangle_2 \otimes |k\rangle_3 \otimes |l\rangle_4$ are the four-mode states. For the first of them, only the passive cavity is in the one-photon state at the initial time, then, the density matrix describing the system can be written as $|1000\rangle\langle 1000|$. The next two cases analyzed correspond to the situation when one of the neutral cavities is in the one-photon state. Finally, for the last situation discussed here, the system starts evolution from the state $|0001\rangle\langle 0001|$. For such a case, only the active cavity is in the one-photon state at the initial time.

3. The Entanglement Generation

In further considerations, we discuss the generation of two-mode entangled states. For the analysis of the degree of bipartite entanglement between two cavities, we apply the entanglement measure which is based on the positive partial transposition criterion, the *bipartite negativity*. It was defined in [42,43] as

$$N_{ij}(\rho_{ij}) = \frac{1}{2}\sum_i |\lambda_n| - \lambda_n, \tag{8}$$

where $\rho_{ij} = Tr_{k,l}\left(\rho_{ijkl}\right)$ is the two-mode density matrix, λ_n is n-th eigenvalue of the matrix $\rho_{ij}^{T_i}$, and $\rho_{ij}^{T_i}$ describes the partial transposition (with respect to the i-th mode) of the two-mode density matrix ρ_{ij}.

In our consideration, we analyze only maximal values of the bipartite negativity N_{0110} defined in the space of four two-mode states: $|0\rangle|0\rangle$; $|0\rangle|1\rangle$; $|1\rangle|0\rangle$; and $|1\rangle|1\rangle$. Therefore, to quantify the bipartite entanglement, we chose the negativity because this quantity can clearly differentiate between entangled and unentangled states when it is applied to two-qubit or qubit—qutrit systems. The negativity takes values between 0 for separable states and 1 for maximally entangled ones. To find values of N_{0110}, we generate the time evolution of the density matrix for the whole system, next, we find the reduced density matrix ρ_{ij} and calculate negativities for the subsystems spanned in the four states. We note that for the system described by $\mathcal{PT}$-symmetric Hamiltonian, its evolution is nonunitary. Therefore, we should normalize the density matrix during the all evolution-time, and then, calculate the negativity with the application of such normalized density matrix ρ_{ijkl}.

As it was mentioned before, we study the possibility of generation of the two-mode entangled state for four cases related to the four initial states (see Figure 3).

First, we consider the situation when the first cavity (passive cavity) is excited, and the initial state is $\hat{\rho}(t=0) = |1000\rangle\langle 1000|$. In Figure 3a, we show the dependence of the maximal values of negativities N_{0110} on γ/β. One can see that only entanglement between modes 1-2 can be produced for all situations considered here. What is relevant is that the degree of entanglement appearing in the system strongly depends on the value of the gain/loss parameter. With the increased value of parameter γ, we can observe decreasing entanglement in all pairs modes. For example, when $\gamma = 0.99\beta$, the maximal value of N_{12} becomes closed to 0.14. For small values of γ/β, the entanglement between all pairs of cavities is generated. Whereas for large values of γ/β, the entanglement corresponding to pairs of cavities 1-3, 1-4, 2-3, 2-4, 3-4 is not produced.

Next, we check the system's ability to produce entanglement when the evolution of our system starts from the state $\hat{\rho}(t=0) = |0100\rangle\langle 0100|$. For such a case, the excitation initially appears in the second cavity (the cavity without loss and gain). In Figure 3b, we see that analogously as in the previous case, the entanglement between all cavities is generated only for small values of γ/β and the strength of maximal possible bipartite entanglement depends on the gain/loss rate again. For weak losses/gains ($\gamma < 0.03\beta$), the entanglement between the cavities 1-2, 2-3, 1-3 is significant, but the strongest entanglement can be observed between subsystems 2 and 3. For strong losses/gain, the entanglement between the cavities 1-2, 2-3 plays the main role, whereas the entanglement between modes 3-4, 2-4, 1-4 is not produced.

In the next analyzed case, for $t = 0$, the third cavity is in the one-photon state and the evolution of the system starts from the state $\hat{\rho}(t=0) = |0010\rangle\langle 0010|$. Figure 3c shows that with increasing parameter γ, all bipartite entanglements become weaker and entanglement between subsystems 1-4 and 2-4 disappears. For small values of γ/β, when the strength of the loss/gain rate increases, the maximal values of all negativities significantly decrease. Whereas for greater values of γ, the values of negativities describing entanglement between modes 1-2, 2-3, 1-3 practically remain constant. For $\gamma > 0.2\beta$, the strongest entanglement we observe is between two neutral cavities.

In the last case, the system's evolution starts from state $\hat{\rho}(t=0) = |0001\rangle\langle 0001|$, and the excitation initially appears in the active cavity. In Figure 3d, we see that analogously as in the previous case, with increasing γ the maximal values of all negativities decrease. What is interesting is that in such a case, all negativities reach nonzero values. The entanglement between cavities 3 and 4 plays the main role. For subsystems 1-2, 2-3, 1-3, 2-4, the entanglement is weaker, and the weakest correlation we observe is between passive and active cavities 1-4. For small values of γ, the negativity N_{14} significantly decreases; and for $\gamma > 0.2\beta$ reaches values close to zero ($N_{14} < 0.01$).

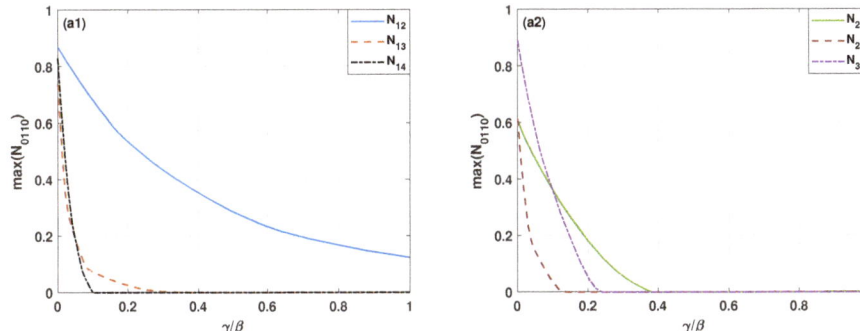

Figure 3. Cont.

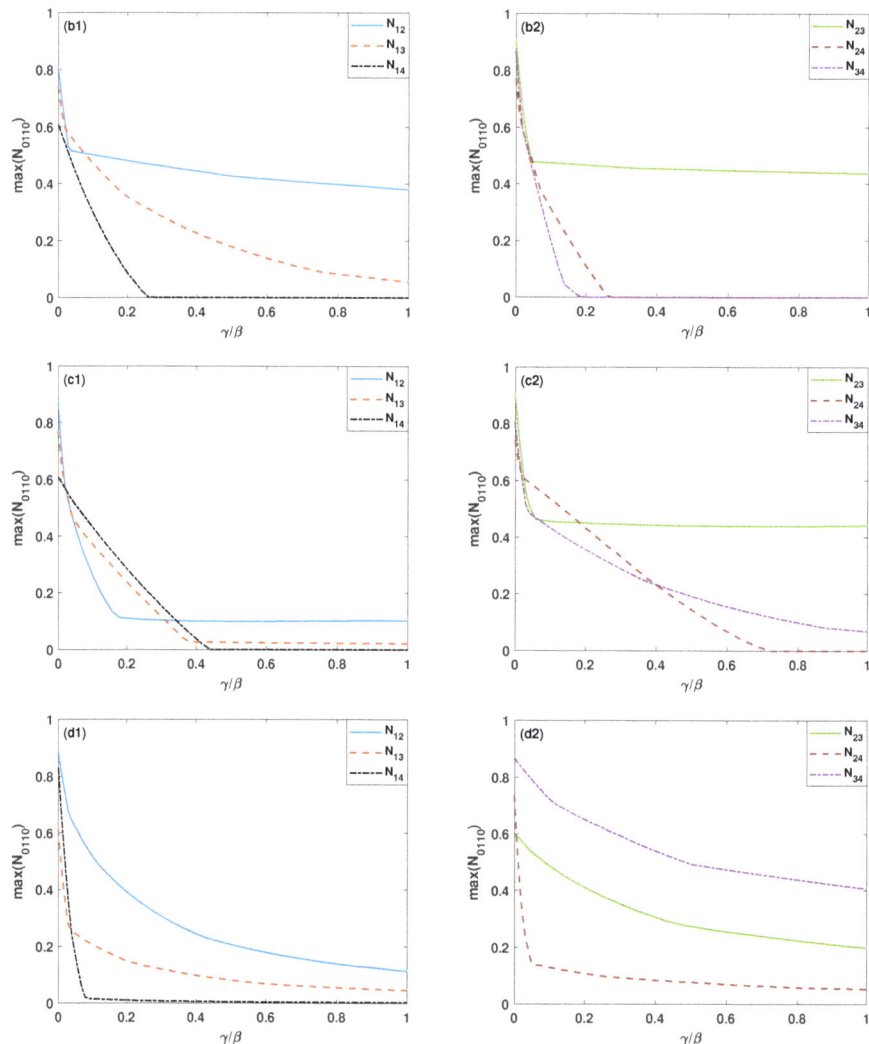

Figure 3. Maximal values of the negativity versus γ/β for $\omega = 2\beta$ and for various initial states: (a) $|1000\rangle\langle 1000|$, (b) $|0100\rangle\langle 0100|$, (c) $|0010\rangle\langle 0010|$, (d) $|0001\rangle\langle 0001|$.

4. Conclusions

Obtaining an entanglement quantum state plays a crucial role in the quantum information theory. The entangled states have many possible applications, such as teleportation and dense coding. Quantum correlations, including entanglement, play a significant role in the development of quantum computation and quantum information processing. Therefore, it is important to understand the entanglement nature of quantum systems. In recent years, a lot of papers have been presented on characterizing entanglement in bipartite systems. The entanglement in such systems can be analyzed relatively easily. The quantum correlations in systems consisting of more than two subsystems have not been understood well, so they are extensively studied. Therefore, we were interested in studying the ability of a $\mathcal{PT}$-symmetric system to generate the entangled states, which are especially interesting from the point of view of quantum information theory. The purpose of this article was to contribute to studies by addressing the entanglement generation.

In this paper, the optical quadrimer system containing four cavities characterized by frequency ω was discussed. The cavities were coupled with each other in such a way that the system formed a chain. Additionally, the first and the last cavity were a subject of losses and gains of energy, respectively, and the rate of losses and gain was balanced.

For such a system, we have found a phase-transition point, and we have shown that when the gain/loss rate is equal or smaller than coupling parameter, all eigenvalues of the system's Hamiltonian are real. In other words, the system is $\mathcal{PT}$-symmetric, contrary to the situation when the gain/loss rate reaches values greater than that of coupling strength and the optical quadrimer is in the broken $\mathcal{PT}$-symmetric phase. For such a case, the system has complex eigenvalue spectra.

We have analyzed four situations corresponding to the four various initial states of the system. We were interested in the possibility of generation of entangled states and the influence of the gain/loss rate strength on the effectiveness of production of such states. We have shown that the degree of entanglement strongly depends on the value of the parameter γ. We have estimated the range of such values for which the strongest bipartite entanglement can be created. For all analyzed cases, when the values of loss/gain rate are small, with increasing γ, the entanglements significantly decrease. For the greater values of loss/gain rate, if the initially active or passive cavity is excited, the strongest entanglement appears between the cavity which, for $t = 0$, is in one-photon state and its nearest neighbor. For the initial time, when one of the neutral cavities is excited, we can observe the strongest entanglement between subsystems 2 and 3.

Funding: The results described in the present work were achieved with the financial support of the ERDF/ESF project "Nanotechnologies for Future" (CZ.02.1.01/0.0/0.0/16_019/0000754) and the program of the Polish Minister of Science and Higher Education under the name "Regional Initiative of Excellence" in 2019–2022, project no. 003/RID/2018/19, funding amount 11 936 596.10 PLN.

Conflicts of Interest: The author declare no conflict of interest. The funders had no role in the design of the study; in the collection, analyses, or interpretation of data; in the writing of the manuscript; or in the decision to publish the results.

References

1. Bender, C.M.; Boettcher, S. Real Spectra in Non-Hermitian Hamiltonians Having $\mathcal{PT}$ Symmetry. *Phys. Rev. Lett.* **1998**, *80*, 5243–5246. [CrossRef]
2. Bender, C.M.; Brody, D.C.; Jones, H.F. Complex Extension of Quantum Mechanics. *Phys. Rev. Lett.* **2002**, *89*, 270401. [CrossRef] [PubMed]
3. Bender, C.M. Making sense of non-Hermitian Hamiltonians. *Rep. Prog. Phys.* **2007**, *70*, 947. [CrossRef]
4. El-Ganainy, R.; Makris, K.G.; Christodoulides, D.N.; Musslimani, Z.H. Theory of coupled optical $\mathcal{PT}$-symmetric structures. *Opt. Lett.* **2007**, *32*, 2632–2634. [CrossRef] [PubMed]
5. Ramezani, H.; Schindler, J.; Ellis, F.M.; Günther, U.; Kottos, T. Bypassing the bandwidth theorem with $\mathcal{PT}$ symmetry. *Phys. Rev. A* **2012**, *85*, 062122. [CrossRef]
6. Miri, M.A.; Regensburger, A.; Peschel, U.; Christodoulides, D.N. Optical mesh lattices with $\mathcal{PT}$ symmetry. *Phys. Rev. A* **2012**, *86*, 023807. [CrossRef]
7. Graefe, E.M.; Jones, H.F. $\mathcal{PT}$-symmetric sinusoidal optical lattices at the symmetry-breaking threshold. *Phys. Rev. A* **2011**, *84*, 013818. [CrossRef]
8. Tchodimou, C.; Djorwe, P.; Nana Engo, S.G. Distant entanglement enhanced in $\mathcal{PT}$-symmetric optomechanics. *Phys. Rev. A* **2017**, *96*, 033856. [CrossRef]
9. Turitsyna, E.G.; Shadrivov, I.V.; Kivshar, Y.S. Guided modes in non-Hermitian optical waveguides. *Phys. Rev. A* **2017**, *96*, 033824. [CrossRef]
10. Zhang, L.L.; Gong, W.J. Transport properties in a non-Hermitian triple-quantum-dot structure. *Phys. Rev. A* **2017**, *95*, 062123. [CrossRef]
11. Mostafazadeh, A. Invisibility and $\mathcal{PT}$ symmetry. *Phys. Rev. A* **2013**, *87*, 012103. [CrossRef]

12. Lü, X.Y.; Jing, H.; Ma, J.Y.; Wu, Y. $\mathcal{PT}$-Symmetry-Breaking Chaos in Optomechanics. *Phys. Rev. Lett.* **2015**, *114*, 253601. [CrossRef] [PubMed]
13. Mostafazadeh, A. Pseudo-Hermiticity versus $\mathcal{PT}$ symmetry: the necessary condition for the reality of the spectrum of a non-Hermitian Hamiltonian. *J. Math. Phys.* **2002**, *43*, 205. [CrossRef]
14. Bender, C.M.; Gianfreda, M.; Özdemir, C.K.; Peng, B.; Yang, L. Twofold transition in $\mathcal{PT}$-symmetric coupled oscillators. *Phys. Rev. A* **2013**, *88*, 062111. [CrossRef]
15. El-Ganainy, R.; Khajavikhan, M.; Ge, L. Exceptional points and lasing self-termination in photonic molecules. *Phys. Rev. A* **2014**, *90*, 013802. [CrossRef]
16. Demange, G.; Graefe, E.M. Signatures of three coalescing eigenfunctions. *J. Phys. A Math. Theor.* **2011**, *45*, 025303. [CrossRef]
17. Heiss, W.D.; Cartarius, H.; Wunner, G.; Main, J. Spectral singularities in $\mathcal{PT}$-symmetric Bose-Einstein condensates. *J. Phys. A Math. Theor.* **2013**. *46*, 275307. [CrossRef]
18. Eleuch, H.; Rotter, I. Clustering of exceptional points and dynamical phase transitions. *Phys. Rev. A* **2016**, *93*, 042116. [CrossRef]
19. Pick, A.; Lin, Z.; Jin, W.; Rodriguez, A.W. Enhanced nonlinear frequency conversion and Purcell enhancement at exceptional points. *Phys. Rev. B* **2017**, *96*, 224303. [CrossRef]
20. Lin, Z.; Pick, A.; Lončar, M.; Rodriguez, A.W. Enhanced Spontaneous Emission at Third-Order Dirac Exceptional Points in Inverse-Designed Photonic Crystals. *Phys. Rev. Lett.* **2016**, *117*, 107402. [CrossRef]
21. Eleuch, H.; Rotter, I. Resonances in open quantum systems. *Phys. Rev. A* **2017**, *95*, 022117. [CrossRef]
22. Zhang, Y.; Zhang, Z.; Yuan, J.; Kang, M.; Chen, J. High-order exceptional points in non-Hermitian Moiré lattices. *Front. Phys.* **2019**, *14*, 53603. [CrossRef]
23. Pick, A.; Zhen, B.; Miller, O.D.; Hsu, C.W.; Hernandez, F.; Rodriguez, A.W.; Soljačić, M.; Johnson, S.G. General theory of spontaneous emission near exceptional points. *Opt. Express* **2017**, *25*, 12325–12348. [CrossRef] [PubMed]
24. Regensburger, A.; Bersch, C.; Miri, M.A.; Onishchukov, G.; Christodoulides, D.; Peschel, U. Parity-time synthetic photonic lattices. *Nature* **2012**, *488*, 167–171. [CrossRef] [PubMed]
25. Peng, B.; Özdemir, C.K.; Rotter, S.; Yilmaz, H.; Liertzer, M.; Monifi, F.; Bender, C.M.; Nori, F.; Yang, L. Loss-induced suppression and revival of lasing. *Science* **2014**, *346*, 328–332. [CrossRef] [PubMed]
26. Zhang, M.; Helmerson, K.; You, L. Entanglement and spin squeezing of Bose-Einstein-condensed atoms. *Phys. Rev. A* **2003**, *68*, 043622. [CrossRef]
27. Vidal, J.; Palacios, G.; Aslangul, C. Entanglement dynamics in the Lipkin-Meshkov-Glick model. *Phys. Rev. A* **2004**, *70*, 062304. [CrossRef]
28. Mohamed, A.; Eleuch, H. Non-classical effects in cavity QED containing a nonlinear optical medium and a quantum well: Entanglement and non-Gaussanity. *Eur. Phys. J. D* **2015**, *69*, 191. [CrossRef]
29. Loss, D.; DiVincenzo, D.P. Quantum computation with quantum dots. *Phys. Rev. A* **1998**, *57*, 120–126. [CrossRef]
30. Miranowicz, A.; Özdemir, Ş.K.; Liu, Y.X.; Koashi, M.; Imoto, N.; Hirayama, Y. Generation of maximum spin entanglement induced by a cavity field in quantum-dot systems. *Phys. Rev. A* **2002**, *65*, 062321. [CrossRef]
31. Cirac, J.I.; Zoller, P. Quantum Computations with Cold Trapped Ions. *Phys. Rev. Lett.* **1995**, *74*, 4091–4094. [CrossRef] [PubMed]
32. Kurpas, M.; Dajka, J.; Zipper, E. Entanglement of qubits via a nonlinear resonator. *J. Phys. Condens. Matter* **2009**, *21*, 235602. [CrossRef] [PubMed]
33. Alexanian, M. Dynamical generation of maximally entangled states in two identical cavities. *Phys. Rev. A* **2011**, *84*, 052302. [CrossRef]
34. Almutairi, K.; Tanaś, R.; Ficek, Z. Generating two-photon entangled states in a driven two-atom system. *Phys. Rev. A* **2011**, *84*, 013831. [CrossRef]
35. Owen, E.T.; Dean, M.C.; Barnes, C.H.W. Generation of entanglement between qubits in a one-dimensional harmonic oscillator. *Phys. Rev. A* **2012**, *85*, 022319. [CrossRef]
36. Brida, G.; Chekhova, M.; Genovese, M.; Krivitsky, L. Generation of different Bell states within the spontaneous parametric down-conversion phase-matching bandwidth. *Phys. Rev. A* **2007**, *76*, 053807. [CrossRef]
37. Coto, R.; Orszag, M.; Eremeev, V. Generation and protection of a maximally entangled state between many modes in an optical network with dissipation. *Phys. Rev. A* **2016**, *93*, 062302. [CrossRef]

38. Ficek, Z.; Tanaś, R. Dark periods and revivals of entanglement in a two-qubit system. *Phys. Rev. A* **2006**, *74*, 024304. [CrossRef]
39. Ficek, Z.; Tanaś, R. Delayed sudden birth of entanglement. *Phys. Rev. A* **2008**, *77*, 054301. [CrossRef]
40. Al-Qasimi, A.; James, D.F.V. Sudden death of entanglement at finite temperature. *Phys. Rev. A* **2008**, *77*, 012117. [CrossRef]
41. López, C.E.; Romero, G.; Lastra, F.; Solano, E.; Retamal, J.C. Sudden Birth versus Sudden Death of Entanglement in Multipartite Systems. *Phys. Rev. Lett.* **2008**, *101*, 080503. [CrossRef] [PubMed]
42. Peres, A. Separability Criterion for Density Matrices. *Phys. Rev. Lett.* **1996**, *77*, 1413–1415. [CrossRef] [PubMed]
43. Horodecki, M.; Horodecki, P.; Horodecki, R. Separability of mixed states: Necessary and sufficient conditions. *Phys. Lett. A* **1996**, *223*, 1–8. [CrossRef]

© 2019 by the author. Licensee MDPI, Basel, Switzerland. This article is an open access article distributed under the terms and conditions of the Creative Commons Attribution (CC BY) license (http://creativecommons.org/licenses/by/4.0/).

Article

Single-Qubit Driving Fields and Mathieu Functions

Marco Enríquez [1],*, Alfonso Jaimes-Nájera [2,3] and Francisco Delgado [1]

[1] School of Engineering and Science, Tecnologico de Monterrey, Atizapán, México 52926, Mexico; fdelgado@tec.mx
[2] Centro de Investigación Científica y de Educación Superior de Ensenada, Unidad Monterrey, PIIT Apodaca, Nuevo León 66629, Mexico; ajaimes@inaoep.mx
[3] Instituto Nacional de Astrofísica, Óptica y Electrónica, Apartado Postal 51/216, Puebla 72000, Mexico
* Correspondence: menriquezf@tec.mx

Received: 13 July 2019; Accepted: 7 September 2019; Published: 16 September 2019

Abstract: We report a new family of time-dependent single-qubit radiation fields for which the correspondent evolution operator can be disentangled in an exact way via the Wei–Norman formalism. Such fields are characterized in terms of the Mathieu functions. We show that the regions of stability of the Mathieu functions determine the nature of the driving fields: For parameters in the stable region, the fields are oscillating, being able to be periodic under certain conditions. Whereas, for parameters in the instability region, the fields are pulse-like. In addition, in the stability region, this family admits solutions for evolution loops in quantum control. We obtain some prescriptions to reach such a control effect. Geometric phases in the evolution are also analyzed and discussed.

Keywords: quantum control; Mathieu functions; time-dependent driving fields

1. Introduction

Exactly solvable models are valuable in quantum mechanics even though the number of cases is very limited. In the case of two-level systems (or *qubits*), some solutions are well-known and longstanding [1–3]. However, the interest in the control of these systems has recently increased due to its potential applications in quantum computation and quantum information [4]. Since the goal is to achieve precise control operations, one should deal with time-dependent driving fields in general. In this context, it is worth mentioning that the design of high-fidelity control operations is required [5,6] and some techniques, such as the adiabatic rapid passage, have been proposed [7]. Nevertheless, the analytical solution for an arbitrary driving field is still an open problem.

The dynamics of a system are determined by the correspondent dynamical law, the interactions to which it is subjected, as well as the initial conditions. However, an alternative approach to obtain exact solutions to such a dynamical law is constituted by the so-called inverse techniques in which some aspects of the dynamics are prescribed and then the interactions are found. This scheme has been widely used to deal with the control of two-level systems through time-dependent radiation fields [8–11]. Recently, a method to generate new solutions to the one qubit dynamical problem has been proposed within this framework. By requiring that the time-evolution operator be exactly factorized as a product of independent exponential factors involving only one Lie algebra generator, new families of time-dependent driving fields are obtained [12]. The evolution-operator disentangling problem lies in the Wei–Norman theorem context [13]. In the case of the $su(2)$ algebra, it is well-known that the direct factorization problem is equivalent to solving a parametric oscillator-like equation whose solutions are known in a limited number of cases [14–17]. Thus, in the inverse solution proposed in [12], one departs from certain known solutions of such an equation and then the driving fields are determined. Accordingly, the dynamics is sensitive to the nature of such solutions.

The main purpose of the present paper is twofold. First, we obtain new families of analytically solvable driving fields using the formalism presented in [12]. Within this framework we show that the dynamics of one qubit interacting with such a family is closely related to the Mathieu functions properties. Second, we present an analysis on the correspondent dynamics. The interest in Mathieu functions lies in the fact that they are widely used in several branches of physics. They have emerged throughout physics mainly in systems with elliptic symmetry or those involving periodic or oscillating behavior. Initially, they were found by Mathieu in 1868 when he studied the oscillating modes of an elliptical membrane [18]. Since then, extensive theoretical work has been done studying their mathematical properties [19]. This has permitted the application of Mathieu functions to study diverse kinds of systems such as elliptical optical waveguides [20], propagation of invariant optical beams [21], the dynamics of electrons in periodic lattices [22], spectral singularities in $\mathcal{PT}$ symmetric potentials in quantum mechanics [23], and Aharonov–Bohm oscillations [24], among others.

The present work is organized as follows. In Section 2, the method to generate exactly solvable driving fields is revisited. We discuss in Section 3 the dynamics of a two-level system interacting with a family of precessing fields with oscillating amplitude generated using this formalism. When the dynamics is solved, we can request the condition to get a cyclic evolution for any initial state in the form of an evolution loop [25,26]. In Section 4, we get the prescriptions of such effect in terms of the dynamical parameters warranting the cyclic behavior for any initial state under the dynamics. An additional analysis on the geometric phase behavior is then conducted. Finally, conclusions and some perspectives are presented in Section 5.

2. Analytically Solvable Driving Fields

This Section is devoted to presenting the formalism to construct driving fields for which the time-evolution operator is exactly factorized. We first consider the two-level system ruled by the time-dependent Hamiltonian

$$H_2(t) = \Delta \sigma_0 + V(t)\sigma_+ + \overline{V}(t)\sigma_-, \tag{1}$$

where $\Delta \in \mathbb{R}$ and $\{\sigma_0, \sigma_\pm\}$ are the three generators of the $su(2)$ algebra defined in terms of the three traditional Pauli operators as follows: $\sigma_0 = \frac{1}{2}\sigma_z$, $\sigma_\pm = \frac{1}{2}(\sigma_x \pm i\sigma_y)$. Furthermore, the interaction of the qubit with a classical field is described by the complex-valued function V. The correspondent time-evolution operator U is a solution of the equation

$$i\frac{dU(t)}{dt} = H_2(t) \cdot U(t), \quad U(0) = \mathbb{I}. \tag{2}$$

As the Hamiltonian (1) is a linear combination of the $su(2)$ algebra generators, the Wei–Norman theorem [13] establishes that the time-evolution operator can be written in the form

$$U(t) = e^{\alpha(t)\sigma_+} e^{\Delta f(t)\sigma_0} e^{\beta(t)\sigma_-}, \tag{3}$$

where the factorization functions α, f, and β satisfy the following coupled system of non-linear equations:

$$\begin{aligned}
\alpha' - \alpha \Delta f' - \beta' e^{-\Delta f} &= -iV, \\
\Delta f + 2\alpha \beta' e^{-\Delta f} &= -i\Delta, \\
\beta' e^{\Delta f} &= -i\overline{V},
\end{aligned} \tag{4}$$

with the initial conditions $\alpha(0) = f(0) = \beta(0) = 0$. By developing the $su(2)$ algebra in the exponential operators, it is an easy task to demonstrate that

$$U(t) = \mathbb{I}(\cosh\frac{\Delta f(t)}{2} + \frac{\alpha(t)\beta(t)}{2}e^{-\frac{\Delta f(t)}{2}}) + 2\sigma_0(\sinh\frac{\Delta f(t)}{2} + \frac{\alpha(t)\beta(t)}{2}e^{-\frac{\Delta f(t)}{2}}) \quad (5)$$
$$+\sigma_+\alpha(t)e^{-\frac{\Delta f(t)}{2}} + \sigma_-\beta(t)e^{-\frac{\Delta f(t)}{2}},$$

where the operator $U(t)$ in (5) has linked conditions fulfilled by $\alpha(t), \beta(t)$, and $\Delta f(t)$. In fact, the unitary condition on $U(t)$: $U(t)U^\dagger(t) = \mathbb{I}$ implies that conditions $\alpha(t) = 0, \beta(t) = 0$, and $\Delta f(t)$ are pure imaginary and fulfill together if at least two of them are requested, which can be proved directly.
Analytical solutions of System (4) are found in a limited number of cases [14]. However, in [12], a formalism to generate exact solutions for the disentangling problem has been developed. In such an approach, the radiation field is given by $V(t) = e^{-i\Delta t}\overline{R}(t)$, where the function R reads

$$R(t) = \frac{R_0}{\mu^2(t)}\exp\left[i\lambda\int_0^t \frac{ds}{\mu^2(s)}\right]; \quad (6)$$

here, the real parameter λ is going to be fixed by the initial conditions and the real-valued function μ satisfies the Ermakov equation

$$\mu''(t) + \Omega^2(t)\mu(t) = \frac{\Omega_0^2}{\mu^3(t)}, \quad \Omega_0 = \left[|R_0|^2 + \frac{\lambda^2}{4}\right]^{1/2}. \quad (7)$$

For reasons to be clarified later, the real number Ω_0 is called the generalized Rabi frequency. Furthermore, the initial conditions are determined by the logarithm derivative of (6) at $t = 0$

$$\frac{R'_0}{R_0} = i\frac{\lambda}{\mu_0^2} - 2\frac{\mu'_0}{\mu_0}, \quad (8)$$

where $R'_0 := R'(0), \mu'_0 = \mu'(0)$ and without loss of generality we take $\mu_0 := \mu(0) = 1$. Thus, once $[\ln R(0)]'$ and R_0 are provided as free parameters, the initial conditions are established. Indeed, given $\delta_1, \delta_2 \in \mathbb{R}$ such that $(\ln R_0)' = \delta_1 + i\delta_2$, it is found that $\lambda = \delta_2$ and $\mu'_0 = -\delta_1/2$. It is a well-known fact that a particular solution to Equation (7) is related to the parametric-like equation [27,28]:

$$\varphi''(t) + \Omega^2(t)\varphi(t) = 0, \quad (9)$$

where Ω^2 is a real-valued function and the initial conditions read

$$\lim_{t\to 0}\frac{1}{R(t)}\left[\frac{\varphi'(t)}{\varphi(t)} + \frac{1}{2}\frac{R'(t)}{R(t)}\right] = 0, \quad \varphi(0) = 1. \quad (10)$$

The former expression is equivalent to the initial condition $\alpha(0) = 0$ (see [14] for details). According to Pinney (alternatively Ermakov), if u and v are two linear independent solutions of (9), then

$$\mu(t) = [u^2(t) - \Omega_0^2 W^{-2}v^2(t)]^{1/2} \quad (11)$$

is a solution of (7). Furthermore, $u(t_0) = \mu(t_0), u'(t_0) = \mu'(t_0), v(t_0) = 0, v'(t_0) \neq 0$, and W stands for the Wronskian of u and v. In addition, note that $R(0) = R_0$ is also a free parameter to be specified and the function φ solves the factorization problem [12,14]. Indeed,

$$\alpha(t) = \frac{i}{R_0}\mu^2(t)\exp\left[-i\Delta t - i\lambda\int_0^t\frac{ds}{\mu^2(s)}\right]\left[\frac{\varphi'(t)}{\varphi(t)} - \frac{\mu'(t)}{\mu(t)} + \frac{i\lambda}{2\mu^2(t)}\right], \quad (12)$$

$$\beta(t) = -iR_0 \int_0^t \frac{ds}{\varphi^2(s)}, \tag{13}$$

$$\Delta f(t) = \ln\left[\frac{\mu^2(t)}{\varphi^2(t)}\right] - i\lambda \int_0^t \frac{ds}{\mu^2(s)} - i\Delta t. \tag{14}$$

The time-evolution of a state can be straightforwardly computed. Because an instance if the qubit's initial state is $|0\rangle$, at any time the state of the system reads

$$|\psi(t)\rangle = e^{-\Delta f(t)/2}[e^{\Delta f(t)} + \alpha(t)\beta(t)]|0\rangle + e^{-\Delta f(t)/2}\beta(t)|1\rangle. \tag{15}$$

An important physical observable is the atomic population inversion defined as $P(t) := \langle \sigma_0 \rangle$. For a general state $|\psi(t)\rangle = c_0|0\rangle + c_1|1\rangle$, the population inversion reads $P(t) = |c_0|^2 - |c_1|^2$. This quantity is defined on the interval $[-1, 1]$. In fact, for the largest (lowest) possible value of $P(t)$, the state is in the excited (ground) state with certainty. In general, for positive (negative) values of the population inversion the probability of finding the state in the upper (lower) energy state is larger. In the Rabi model, the population inversion is a periodic function of time where field amplitude and detuning (the gap between the field frequency and the spacing level energy of the atom) determine the oscillation period [12,14,29].

We finish this section by summarizing the method: One should start with the second-order differential Equation (9) choosing Ω such that the solutions are known. The function μ is then obtained via the Ermakov–Pinney solution and so it is possible to generate families of analytically solvable driving fields (6) for which the initial conditions constitute free parameters to control the dynamics. Some examples have been reported in [12]. For instance, if Ω^2 is a negative real constant, the correspondent driving field is a decaying one. However, a precessing field with oscillating amplitude is achieved if Ω^2 is a positive real constant Ω_1. Such a family of control fields constitutes a generalization of the circularly polarized field, which is obtained as a particular case of this model.

3. Dynamics in a Precessing Field with Oscillating Amplitude

In this section, we report a new family of driving fields for which the time-evolution operator is exactly disentangled. We consider the function $\Omega^2(t) = \omega_0 - \omega_1 \cos^2(t)$ in the parametric oscillator-like Equation (9), which can be written as a Mathieu equation [19,30] as follows:

$$\varphi''(t) + [a - 2q\cos(2t)]\varphi(t) = 0, \tag{16}$$

where the identity $2\cos^2 t = 1 + \cos(2t)$ has been used and

$$a = \omega_0 - \frac{1}{2}\omega_1, \qquad q = \frac{1}{4}\omega_1. \tag{17}$$

Furthermore, the following initial conditions read $[\ln R(0)]' = i\delta$, where $\delta \in \mathbb{R}$ and $\varphi_0 = 1$. According to (11), the function μ can be written as

$$\mu^2(t) = u^2(t) + \left(\frac{\Omega_0 v(t)}{W}\right)^2, \tag{18}$$

where $\Omega_0^2 = |g|^2 + \delta^2/4$ has been determined considering $R_0 = -i\bar{g}$, $g \in \mathbb{C}$. Note that these particular expressions for the initial conditions retrieve the well-known Rabi frequency in the problem of a qubit interacting with a circularly polarized field. Furthermore, it is easy to prove that the Wronskian W does not depend on t.

In most cases, the two linearly independent solutions of the Mathieu Equation (16) can be written in terms of the even and odd Mathieu functions $M_C(a, q, t)$ and $M_S(a, q, t)$ [19]. As is clarified in the next section, we restrict ourselves to this case.

3.1. The Dynamics of Driving Fields and the Theory of Mathieu Functions

In this section, we study the behavior of the solution μ (18) of the Ermakov Equation (7) as a function of t, but also as a function of its parameters. The behavior of the driving fields can be very different depending on the values of the frequencies ω_0 and ω_1. Moreover, $R(t)$ can be a periodic, non-periodic, oscillating, or pulsed-like function of time. In this respect, the theory of Mathieu functions will be useful.

Given a value of the parameter q of the Mathieu Equation (16), there exist values of a for which the Mathieu functions can be periodic, bounded, or unbounded [19]. When the Mathieu function M_C (M_S) is periodic, the values of a are called *characteristic values* and they are denoted by a_r (b_r), becoming functions of the parameter q

$$a_r = r^2 + \sum_{i=1}^{\infty} \alpha_i(r) q^i, \qquad b_r = r^2 + \sum_{i=1}^{\infty} \beta_i(r) q^i, \qquad (19)$$

where $\alpha_i(r), \beta_i(r) \in \mathbb{R}$ are given in [19]. The parameter $r \geq 0$ is called the *characteristic exponent* and it determines the periodic properties of the Mathieu functions. The even Mathieu function $M_C(a_r, q, t)$ is periodic with period π or 2π if the characteristic exponent r can be written as $2n$, or $2n+1$, respectively, where $n \in \mathbb{N} \cup \{0\}$. The odd Mathieu function $M_S(b_r, q, t)$ is periodic with period π or 2π if r can be written as $2n+2$, or $2n+1$, respectively [19]. When r is an integer $a_r \neq b_r$, then the Mathieu functions $M_C(a_r, q, t)$ and $M_S(b_r, q, t)$ are not solutions to the same Mathieu equation. On the other hand, when r is not an integer, $a_r = b_r$ holds. Then, $M_C(a_r, q, t)$ and $M_S(a_r, q, t)$ are solutions to the same Mathieu equation, and moreover, they are linearly independent [19]. Considering r as a positive rational number, then, it can be written as $r = m + p/s$, where $m \geq 0$ is an integer and $p \neq 0$, s are relative prime integers. Hence, Mathieu functions $M_C(a_r, q, t)$ and $M_S(a_r, q, t)$ are periodic with period $2\pi s$, for $s > 2$. On the other hand, if r is an irrational number, Mathieu functions are not periodic but bounded functions which do not decay to zero as $t \to \infty$ [19].

We define the $q - a$ space as the two dimensional space in which q and a are the abscissa and ordinate, respectively. The behavior of the Mathieu functions is dictated by the region of the $q - a$ space in which the point (a, q) lies, and so the discussion of the previous paragraph can be summarized into a picture in the $q - a$ space as follows. As mentioned above, the characteristic values are functions of q, therefore in the $q - a$ space, the curves of a_r and b_r lie there as functions of q. Such curves are from now on called the *characteristic curves*. In Figure 1a, the behavior of the characteristic curves can be observed in the space $q - a$ for different integer values of r, while in Figure 1b, the case is shown with several non-integer values of r. The characteristic curves for all non-integer positive values of r, a case in which $a_r = b_r$, saturate a certain region of the $q - a$ space [19] which is shown shaded in Figure 1b. This region, along with the characteristic curves of integer values of r, is known as the region of *stability* of the Mathieu functions. A Mathieu function is called *stable* if it is bounded or tends to zero as $t \to +\infty$, as well as it is called *unstable* if it diverges as $t \to +\infty$. Therefore, if the point (a, q) lies inside (outside) the region of stability, the corresponding Mathieu function is a bounded (unbounded) function of t. Regarding Equation (17), it can be observed that ω_0 can be written as $\omega_0 = a + \omega_1/2$. Taking a as a characteristic value $a_r(q)$, ω_0 can be considered as a characteristic frequency (value), in the sense that

$$\omega_0(\omega_1, r) = a_r(\omega_1) + \omega_1/2, \qquad (20)$$

in the analog $\omega_1 - \omega_0$ space, since $q = \omega_1/4$. Figure 2 shows the behavior of the characteristic curves $\omega_0(\omega_1, r)$ in the $\omega_1 - \omega_0$ space.

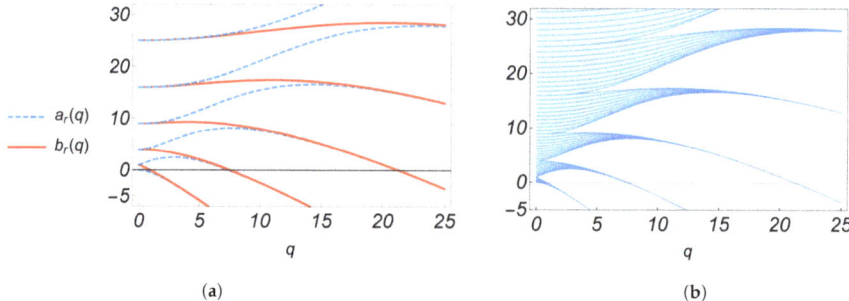

Figure 1. (a) The characteristic values $a_r(q)$ (blue-dashed curves) for $r = 0, 1, 2, 3, 4, 5$, and b_r (red curves) for $r = 1, 2, 3, 4, 5$, as functions of q, for several values of $r \in \mathbb{N}$. Note that $a_r(0) = b_r(0) = r^2$. (b) The shaded region is produced by all the characteristic curves with non-integer characteristic values r, case in which $a_r = b_r$.

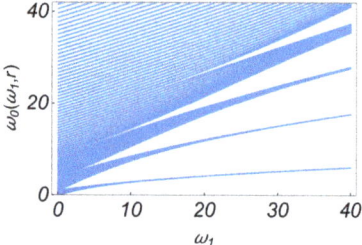

Figure 2. The characteristic frequency $\omega_0(\omega_1, r) = a_r(\omega_1) + \omega_1/2$ as functions of ω_1, for several rational values of r.

Nevertheless, we are interested in the two linearly independent solutions of the Mathieu equation conforming to the Ermakov equation solution μ. Then, we must separate the $\omega_1 - \omega_0$ space into two regions: the one in which *both* linearly independent solutions are stable, which will be denoted as $\mathcal{A}$; and the one in which *at least* one of the two solutions is unstable, represented by $\mathcal{A}^C$. In region $\mathcal{A}$, the function μ is a bounded function of t, while in the region $\mathcal{A}^C$, μ is unbounded, as can be observed from Equation (18).

As mentioned above, within the stability region, there exist two cases: (*i*) the characteristic exponent r is not an integer; and (*ii*) it is an integer. In case (*i*), both Mathieu functions M_C and M_S are linearly independent stable solutions of the Mathieu Equation (16). In case (*ii*), the second linearly independent solution M_C as well as the second linearly independent M_S are unstable [19]. On the other hand, in the instability region, both linearly independent solutions of the Mathieu equation are unbounded. Therefore, $\mathcal{A}$ can be described as the region of the $\omega_1 - \omega_0$ space in which the characteristic values r are positive non-integers (see Figure 2), namely

$$\mathcal{A} = \{(\omega_1, \omega_0) \in \mathbb{R}^2 \mid \omega_0(\omega_1, r) = a_r(\omega_1) + \omega_1/2, r > 0, r \notin \mathbb{Z}\}, \qquad (21)$$

where $a_r(\omega_1)$ is the characteristic value as function of $\omega_1(q)$, see Equation (17). Let us denote by $\mathcal{B}$ the set of points of the characteristic curves for integer values of $r > 0$ (denoted in Figure 1a), so the stability region in the $\omega_1 - \omega_0$ space can be written as $\mathcal{A} \cup \mathcal{B}$ (note that the set $\mathcal{B}$ is part of the boundary of $\mathcal{A}$ (see Figure 1)). Since this is case (*ii*), one of the solutions to the Mathieu equation is unstable, then $\mathcal{B} \subset \mathcal{A}^C$. Hence, within $\mathcal{A}^C$, there are two cases: the point (ω_1, ω_0) belongs to the unstable region or to $\mathcal{B}$.

For simplicity, we focus on the case in which μ can be written in terms of the Mathieu functions M_C and M_S. This case includes the whole region $\mathcal{A}$ as well as $\mathcal{A}^C - \mathcal{B}$ (unstable region case), that is to say, we are excluding the points in $\mathcal{B}$. Nevertheless, as the second solutions in $\mathcal{B}$ are unstable, the asymptotic behavior of μ for $t \to \infty$ for $(\omega_1, \omega_0) \in \mathcal{A}^C$ is similar whether (ω_1, ω_0) belongs to $\mathcal{B}$ or not.

Since Equation (16) is invariant under a complex conjugate operation the real or imaginary part of any of its solutions is also a solution and being μ a real-valued function, it can be written as

$$\mu^2(t) = c^2 \text{Re}[M_C(t)]^2 + \frac{\Omega_0^2}{W^2} \text{Re}[M_S(t)]^2, \tag{22}$$

where we have defined $u(t) = c\text{Re}[M_C(t)]$, $c = 1/\text{Re}[M_C(0)]$, and $v(t) = \text{Re}[M_S(t)]$ (the dependence on the frequencies ω_0, ω_1 has been omitted for shortness) in order to satisfy the initial conditions stated in Equation (10) and the ones mentioned after Equation (18). Furthermore, W denotes the Wronskian of two functions. Then, the driving field $R(t)$ can be written as

$$R(t) = \frac{-i\bar{g}}{c^2 \text{Re}[M_C(t)]^2 + \frac{\Omega_0^2}{W^2} \text{Re}[M_S(t)]^2} \exp\left(i\delta \int_0^t \frac{ds}{c^2 \text{Re}[M_C(s)]^2 + \frac{\Omega_0^2}{W^2} \text{Re}[M_S(s)]^2}\right). \tag{23}$$

In order to write the factorizing functions, we have to obtain the solution to the parametric oscillator-like Equation (16), satisfying the initial conditions (8), which can be written as

$$\varphi(t) = \frac{1}{M_C(0)} M_C(t) - \frac{i\delta}{2M_S'(0)} M_S(t), \tag{24}$$

where $M_S'(t) = dM_S(t)/dt$. The factorizing functions can be written as

$$\alpha(t) = \frac{\mu^2(t)}{-i\bar{g}} \left[\frac{\varphi'(t)}{\varphi(t)} - \frac{\mu'(t)}{\mu(t)} + \frac{i\delta}{2\mu^2(t)}\right] e^{-i(\Delta t + \mu_1(t))}, \tag{25}$$

$$\beta(t) = R_0 \int_0^t \frac{ds}{\varphi^2(s)}, \tag{26}$$

$$\Delta f(t) = \ln\left[\frac{\mu^2(t)}{\varphi^2(t)}\right] - i\mu_1(t) - i\Delta t, \tag{27}$$

where

$$\mu_1(t) = \delta \int_0^t \frac{ds}{\mu^2(s)}. \tag{28}$$

Now, we are able to calculate the time-evolution of the initial state $\psi(0) = |1\rangle$ given in (15). As commented previously, population inversion, $P(t) \in [-1, 1]$ is a physical quantity very useful and descriptive for two-level systems with respect to the system dynamics. It depicts, for the quantum state, the dynamical variation between the excited state $|0\rangle$ ($P(t) = 1$) and the base state $|1\rangle$ ($P(t) = -1$). Thus, in our case, the population inversion can be written after some algebra as

$$P(t) = \mu^2(t) |\varphi(t) \beta(t)|^2 \left\{ \left|\frac{1}{\beta(t)\varphi^2(t)} + \frac{\alpha(t)}{\mu^2(t)} e^{i[\Delta t + \mu_1(t)]}\right|^2 - \frac{1}{\mu^4(t)} \right\}. \tag{29}$$

3.1.1. Driving Fields in the Region $\mathcal{A}$

In this case, the solution μ to Ermakov equation is a time oscillating function that can be periodic if the characteristic exponent r is a rational number. Hence, the driving field $R(t)$, as well as V, is an oscillating function which becomes periodic when the parameters g, δ, Δ, and the characteristic exponent r satisfy the condition [12]

$$p\delta \int_0^\tau \frac{ds}{\mu^2(s)} + \Delta\tau \equiv 0 \quad (\text{mod } 2\pi), \tag{30}$$

where p is a natural number and τ is the period of μ. Solutions exhibit lots of possibilities due to the set of physical parameters a, q, g, δ, Δ, and τ (as well as ω_1, ω_2 instead of a, q). They could be selected or highlighted to get several possible control effects: periodicity in the field $R(t)$, the reaching of evolution loops, or simply the control of population inversion. We analyze several of these effects separately in the following sections of the article.

Figure 3 shows three examples of periodic driving fields and their associated population inversion. In these cases, the characteristic exponent r is a rational number, so the associated μ is periodic and the parameters g and δ are such that $R(t)$ is a periodic function of time. Furthermore, in order to show the time flow correspondence between the left plots with the right ones, a color scale from blue to red has been used. As can be observed from Figure 3b,e, each minimum of the population inversion barely corresponds to a local maximum of the transverse driving field amplitude $|R(t)|$. Furthermore, the non-negative parameter p is related to the number of "petals" in each flower-like structure of the driving field. In [12], it is shown that our model can describe a two-level system interacting with a magnetic transverse field $B_\perp$. It is easy to show that real and imaginary parts of $R(t)$ are nothing other than the field components performing a certain rotation. This fact is due to the transverse field $B_\perp^2 = B_x^2 + B_y^2 = \text{Im}[R(t)]^2 + \text{Re}[R(t)]^2$ causing the state to instantaneously precess around the direction of the entire field, the direction of such precession becomes more horizontal while $|B_\perp| \gg |B_z| = \frac{|\Delta|}{2}$. Thus, when the relative orientation between the field direction and the state direction on the Bloch sphere becomes more perpendicular, the states rotates more extremely on the sphere, being able to reach a more extreme position there, just as it happens in the case of a circularly polarized field [31]. These rotations becomes more effective the bigger the field is because the instantaneous angular frequency depends on it. When the value of the parameter δ is changed, with the others held constant, an oscillatory non-periodic driving field is obtained (see the continuous blue curves in Figure 4). In this case, the changes in the amplitude of the driving field are not fast enough for the population inversion to reach its minimum value -1. Now, by only increasing the parameter g and holding the others fixed (the magnitude of $B_\perp$ is proportional to $|g|$), we observe in Figure 4 (red dashed curves) that the changes in the amplitude of the driving field are increased with respect to the previous case, pushing the increment of the range of the peaks of the population inversion towards $(-1, 1)$. The positions of the maxima of the population inversion remain unchanged. We can conclude, in this case, that as the parameter g increases, the peaks of the population inversion are sharpened, presenting a resonant-like behavior. Furthermore, in any of the previous cases, as shown at the end of Section 3.1, the population inversion is periodic.

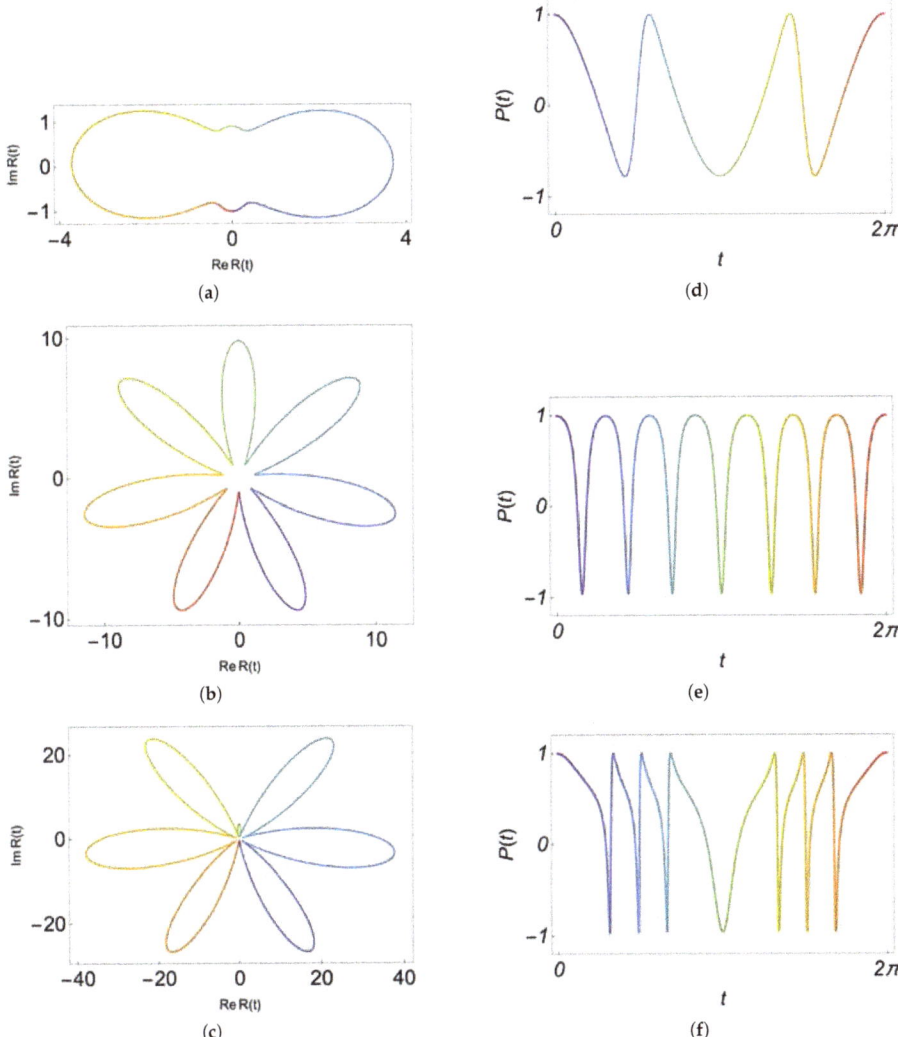

Figure 3. The driving field (23) in the complex plane for $g = 1$, $\Delta = 1$, (**a**) $r = 1.5$, $\omega_1 = 4$, $\omega_0 = 4.53718$, $\delta = 0.7071$, (**b**) $r = 3.5$, $\omega_1 = 4$, $\omega_0 = 14.2946$, $\delta = 0.28867$, (**c**) $r = 3.5$, $\omega_1 = 40$, $\omega_0 = 36.2607$, $\delta = 0.28867$. In (**d**–**f**), the corresponding population inversion is shown. All of the previous values of ω_0 correspond to characteristic values of the Mathieu functions (16) with rational values of r, that is, $(\omega_1, \omega_0) \in \mathcal{A}$. Time flows from blue to red matching the scale between the types of plots.

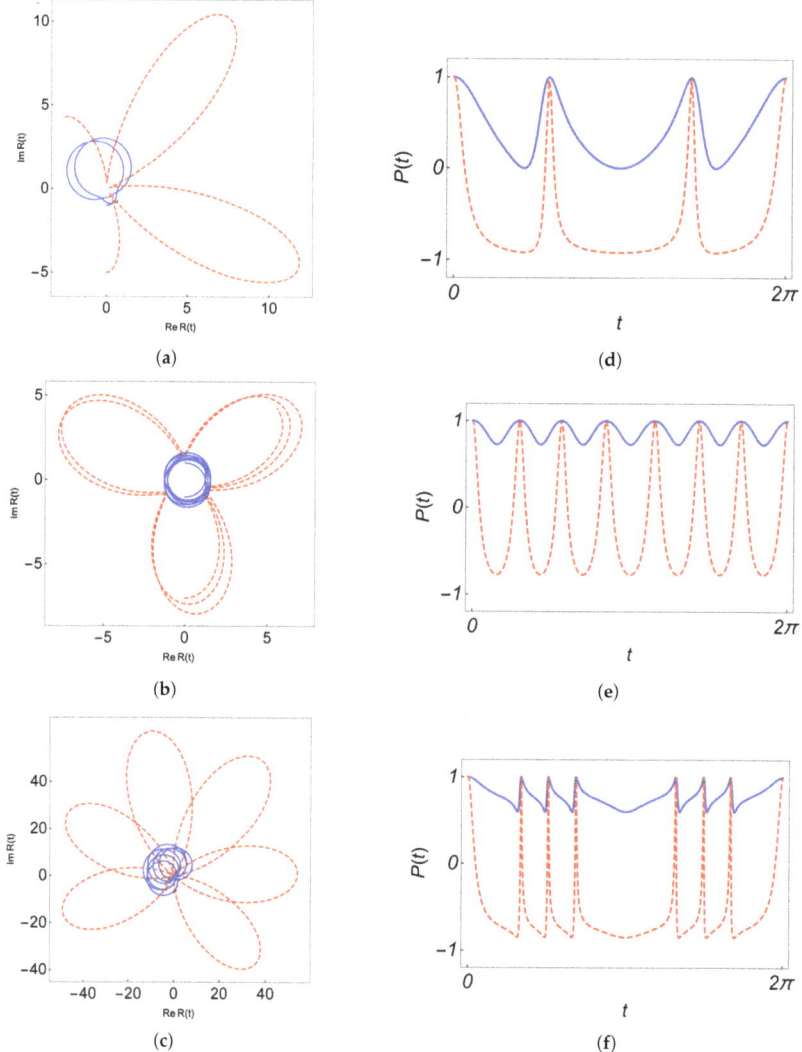

Figure 4. The driving fields (23) in the complex plane (left column) and the associated population inversion (right column) for the cases of Figure 3, except for the values of g and δ, which has been increased as follows: (**a,d**) $g = 1$, $\delta = 2$ (blue), and $g = 5$, $\delta = 2$ (red dashed), (**b,e**) $g = 1$, $\delta = 5$ (blue), and $g = 7$, $\delta = 5$ (red dashed), (**c,f**) $g = 1$, $\delta = 4$ (blue), and $g = 7$, $\delta = 4$ (red dashed). The driving field precesses counterclockwise.

3.1.2. Driving Fields in the Region $\mathcal{A}^C$

As mentioned above, at least one of the Mathieu functions, which conform the solution μ to the Ermakov equation, is an unbounded function diverging as $t \to \infty$. In this case, we have observed that the zeros of the Mathieu functions $M_C(t)$ and $M_S(t)$ tend to approach each other more and more as t increases, a phenomenon known as condensation of zeros [19]. This gives rise to pulsed-like peaks in the corresponding driving field $R(t)$ ($|R(t)| \propto 1/\mu^2$), as can be observed in Figure 5a. Therefore, as time passes, the pulses increase their size and become narrower causing a pulsed-like behavior of the population inversion, as can be observed in Figures 5b and 5c.

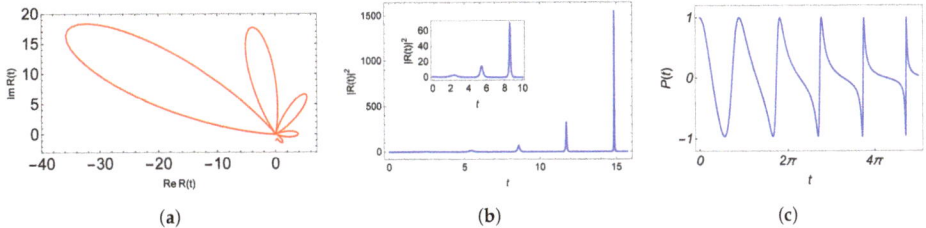

Figure 5. (a) The driving field (23) in the complex plane. (b) Its square modulus as a function of time. The inset is a zooming of the plot for $t \in [0, 10]$ to show the correspondent local maxima, and (c) the associated population inversion with $\omega_1 = 1$, $\omega_0 = 1.5$, $g = 1$, $\delta = 0.2887$, where $(\omega_0, \omega_1) \notin \mathcal{A}$. The driving field precesses counterclockwise.

4. Evolution Loops, Cyclic Evolution and Phases

Evolution loops are a control phenomena present in certain quantum systems admitting dynamic solutions to the restriction $U(\tau) = e^{i\phi}\mathbb{I}$ for some $\tau > 0$, which clearly revert the evolution after time τ. Furthermore, the parameter ϕ is the global phase acquired during the process. It is relevant because that reduction becomes independent from the initial states, thus, any initial state evolves after such time. These phenomena boost other intermediate effects which have value in quantum control. In this section, we show that the field being considered admits this kind of effects. In the current case, because if $\alpha(\tau) = 0$ and $\beta(\tau) = 0$, then $\Delta f(\tau)$ is automatically pure imaginary, then, if in addition, $\phi = \Delta f(\tau) = 4n\pi i, n \in \mathbb{Z}$, $U(\tau)$ reduces exactly to $\mathbb{I}$. Such is the condition to reach evolution loops for some $t = \tau$ (otherwise, if only $\alpha(\tau) = 0 = \beta(\tau)$ are fulfilled, the evolution reaches the initial state with a non-zero phase). Then, the procedure to find such prescriptions is based in the finding of τ satisfying $\alpha(\tau) = 0 = \beta(\tau)$, then by using (27), we can adjust the value of Δ to reduce $\Delta f(t)$ to $4n\pi i, n \in \mathbb{Z}$. This is always possible due to the first two terms on the right side being independent of Δ. Despite this, there are a lot of solutions because there are five free parameters: a, q (otherwise ω_0, ω_1), δ, g, t.

Figure 6 shows three examples of such an effect from a large variation in the initial state $|1\rangle$ until a smaller one in terms of parameters a, q, g, δ, Δ, and τ. Those parameters work in a combined way to give dynamical effects, thus, they are not related with the periodicity of $R(t)$, $P(t)$ of $U(t)$ in an exclusive way. In this sense, examples being considered are only illustrative among a wide variety of possibilities (obtained numerically upon the conditions $\alpha(\tau) = \beta(\tau) = 0$), but exhibiting different ranges of variation for $P(t)$. In any case, note that g and δ play an important role in such variation, as is expected. Plots on the left exhibit the value of $P(t)$ while plots on the right show the dynamics represented on the Bloch sphere. Nevertheless, evolution loops are independent from the initial states, examples consider the initial state as the ground state, $|1\rangle$. The color changes from blue to red in agreement with the $P(t)$ value. Note that the intermediate reaching of unitary multiples of $|1\rangle$ with non-zero phases until the zero phase is reached in τ. The last example in Figure 6c meets the evolution loop under periodic driving field, generating a more regular dynamics around $|1\rangle$.

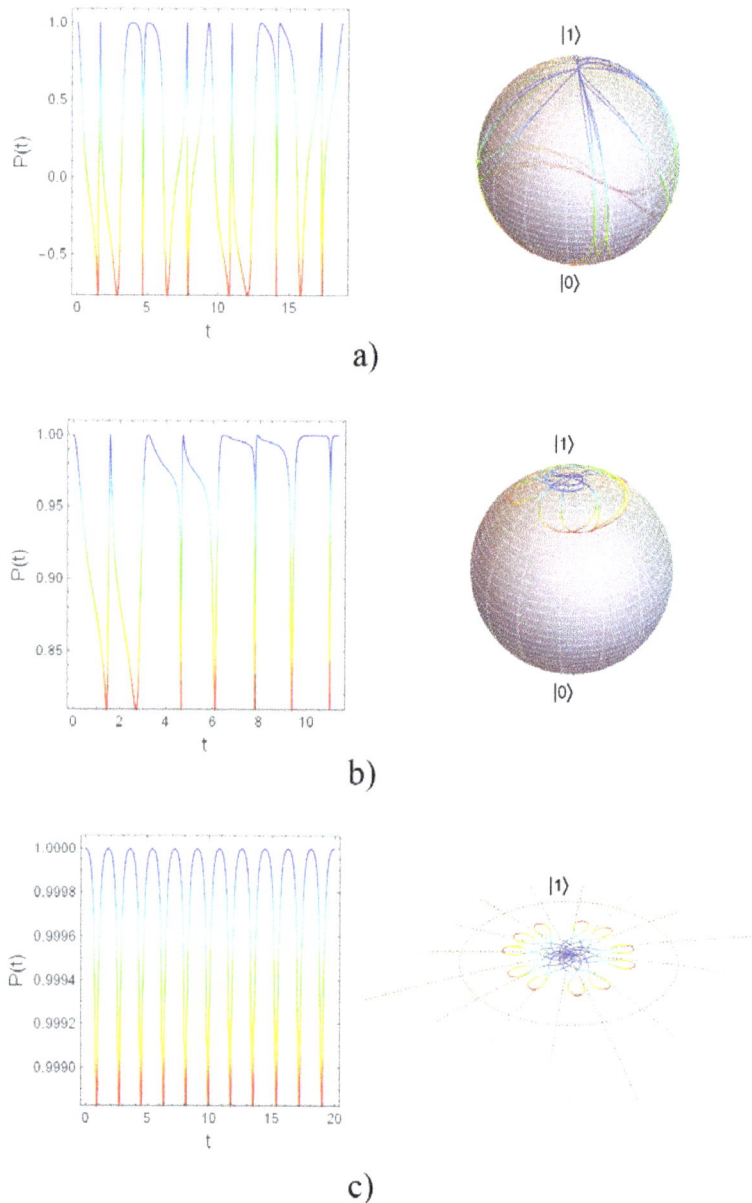

Figure 6. Population inversion together with the dynamics represented on Bloch spheres for several prescriptions generating evolution loops. In any case, the color depicts the value of $P(t)$ in agreement with the left plot. (**a**) $a = 3.37813, q = 2.6118, g = 0.53635, \delta = 3.30558, \Delta = 0.190982, t = 11.3718$. (**b**) $a = 3.13268, q = 2.70972, g = 1.93404, \delta = 1.41305, \Delta = 0.74138, t = 18.7095$. (**c**) $a = 3.01588, q = 0.022235, g = 0.0123357, \delta = 1.0189, \Delta = 0.00100538, t = 18.3134$ depicting a tiny regular evolution loop with a periodic field around of $|1\rangle$.

Dynamical and Geometric Phases

It is well-known that a geometric phase exists related to a quantum system cyclic evolution governed by a slow change of parameters [32,33]. The relevance of such a phase factor lies in the foundations of quantum theory; however, it has also recently found important applications in quantum information and computation, as a part of the system evolution [34,35].

Consider the dynamic phase ϕ_d, as well as the Aharonov–Anandan geometric phase ϕ_g in evolution loop dynamics [36,37], which depicts the phase when the system returns to its initial physical state at the time τ (in the current case to the state $|0\rangle$):

$$\phi_d = -\int_0^\tau \langle \psi(t)|H_2(t)|\psi(t)\rangle dt \quad \to \quad \phi_g = \arg(\langle \psi(0)|\psi(\tau)\rangle) - \phi_d \equiv \phi - \phi_d, \quad (31)$$

which even in the simplest case of a time-independent magnetic field, a non-trivial value is obtained: $\phi_d = -\pi \cos\theta_0$ and $\phi_g = \pi(\cos\theta_0 - 1)$, where θ_0 is the relative angle between the magnetic field and the initial Bloch vector. These values are indeed independent of the field amplitude [38]. Conversely, in time-dependent cases, the dynamical trajectory on the Bloch sphere can be manipulated with respect to the field parameters, yielding different values of the geometric phase. Such is our case, as shown in the examples in the Figure 6: (a) $\phi_d = -38.057, \phi_g = 126.022, \phi = 28\pi$, (b) $\phi_d = -2.19494, \phi_g = 52.460, \phi = 16\pi$, and (c) $\phi_d = -13.4832, \phi_g = 88.881, \phi = 24\pi$. In these examples, the global phase ϕ has been selected as the minimum value assuring the positivity for Δ. The present analysis represents a first step in the construction of logical gates for holonomic quantum computation [34,35]. These results will be reported elsewhere.

5. Conclusions

We have presented the study of driving fields generated by the inverse approach given in [12], departing from a sinusoidal parametric oscillator-like equation. Regarding the solution of the associated Ermakov equation, the driving fields and the population inversion can be written in terms of the Mathieu functions. We have shown that the theory of Mathieu functions is determinant in the dynamical analysis of such driving fields, as a clear split in two regions for the space of the frequencies ω_0 and ω_1 is acquired. Thus, different dynamical behaviors of the driving fields are shown. In the region $\mathcal{A}$, we have shown that the driving fields have an oscillatory nature, but they can still be periodic or non-periodic depending on the value of the parameters g and δ given by a precise prescription obtained analytically. In this same region, the population inversion is periodic provided that the associated Mathieu functions are also periodic. In the complement of the previous region, for example $\mathcal{A}^C$, the behavior of the driving fields is pulse-like; with the amplitude and sharpness of their peaks increasing in time, yielding a resonant-like behavior for the associated population inversion. We consider that this system presents the advantage of having several kinds of dynamics that can be prescribed on demand by fine tuning the frequencies ω_0 and ω_1 (alternatively a or q) of the parametric oscillator-like potential. Together, we showed that evolution loop solutions are numerically achievable and the correspondent geometric phases can be calculated. Our study represents the first stage in the development of technological applications involving single-qubit devices. For instance, in some many-body models, precise selective control operations on single qubits are required, which could be achievable through the driving fields obtained in this work. Such analysis will be reported elsewhere. We hope our results shed some light on the matter.

Author Contributions: Investigation, M.E., A.J.-N. and F.D.; Writing—original draft, M.E., A.J.-N. and F.D.; Writing—review & editing, M.E., A.J.-N. and F.D.

Funding: The support of CONACyT through project A1-S-24569 is acknowledged.

Acknowledgments: M.E. and F.D. would like to acknowledge to School of Engineering and Science of Tecnologico de Monterrey as well as the financial support of Novus Grant with PEP no. PHHT023_17CX00001, TecLabs,

Tecnologico de Monterrey, Mexico, in the production of this work. A.J.-N. thanks the support of INAOE and CICESE.

Conflicts of Interest: The authors declare no conflict of interest.

References

1. Rabi, I. Space quantization in a gyrating magnetic field. *Phys. Rev.* **1937**, *51*, 652. [CrossRef]
2. Zener, C. Non-adiabatic crossing of energy levels. *Proc. R. Soc. A* **1932**, *137*, 696–702. [CrossRef]
3. Schwinger, J. On non-adiabatic processes in inhomogeneous fields. *Phys. Rev.* **1937**, *51*, 648. [CrossRef]
4. Nielsen, M.A.; Chuang, I.L. *Quantum Computation and Quantum Information*; Cambridge University Press: Cambridge, UK, 2000.
5. Zeng, J.; Barnes, E. Fastest pulses that implement dynamically corrected single-qubit phase gates. *Phys. Rev. A* **2018**, *98*, 012301. [CrossRef]
6. Motzoi, F.; Wilhelm, F.K. Improving frequency selection of driven pulses using derivative-based transition suppression. *Phys. Rev. A* **2013**, *88*, 062318. [CrossRef]
7. Stefanatos, D.; Paspalakis, E. Resonant adiabatic rapid passage with only z-field control. *Phys. Rev. A* **2019**, *100*, 012111. [CrossRef]
8. Fernández, D.J.; Rosas-Ortiz, O. Inverse techniques and evolution of spin-1/2. *Phys. Lett. A* **1997**, *236*, 275. [CrossRef]
9. Barnes, E.; Das Sarma, S. Analytically solvable driven time-dependent two-level quantum systems. *Phys. Rev. Lett.* **2012**, *109*, 060401. [CrossRef]
10. Vitanov, N.V. Complete population inversion by a phase jump: An exactly soluble model. *New J. Phys.* **2007**, *9*, 58. [CrossRef]
11. Messina, A.; Nakazato, H. Analytically solvable Hamiltonians for quantum two-level systems and their dynamics. *J. Phys. A Math. Theor.* **2014**, *47*, 445302. [CrossRef]
12. Enríquez, M.; Cruz, S.C.y. Exactly Solvable One-Qubit Driving Fields Generated via Nonlinear Equations. *Symmetry* **2018**, *10*, 567. [CrossRef]
13. Wei, J.; Norman, E. Lie algebraic solution of linear differential equations. *J. Math. Phys.* **1963**, *4*, 575–581. [CrossRef]
14. Enríquez, M.; Cruz, S.C.y. Disentangling the time-evolution operator of a single qubit. *J. Phys. Conf. Ser.* **2017**, *839*, 012015. [CrossRef]
15. Datolli, G.; Solimeno, S.; Torre, A. Algebraic time-ordering technics and harmonic oscillator with time-dependent frequency. *Phys. Rev. A* **1986**, *34*, 2646. [CrossRef] [PubMed]
16. Datolli, G.; Torre, A. SU(2) and SU(1,1) time-ordering theorems and Bloch-type equations. *J. Math. Phys.* **1987**, *28*, 618. [CrossRef]
17. Datolli, G.; Richetta, M.; Torre, A. Evolution of SU(2) and SU(1,1) states: A further mathematical analysis. *J. Math. Phys.* **1988**, *29*, 2586. [CrossRef]
18. Mathieu, É. Le mouvement vibratoire d'une membrane de forme elliptique. *J. Math. Pures Appl.* **1868**, *13*, 137–203.
19. McLachlan, N.W. *Theory and Application of Mathieu Functions*; Oxford University Press: London, UK, 1951.
20. Li, S.; Wang, B.S. Field expressions and patterns in elliptical waveguides. *IEEE Trans. Microw. Theory Tech.* **2000**, *48*, 864–867.
21. Gutiérrez-Vega, J.C.; Iturbe-Castillo, M.D.; Chávez-Cerda, S. Alternative formulation for invariant optical fields: Mathieu beams. *Opt. Lett.* **2000**, *25*, 1493–1495. [CrossRef]
22. Carver, T.R. Mathieu's functions and electrons in a periodic lattice. *Am. J. Phys.* **1971**, *39*, 1225–1231. [CrossRef]
23. Sinha, A.; Roychoudhury, R. Spectral singularity in confined $\mathcal{PT}$ symmetric optical potential. *J. Math. Phys.* **2013**, *54*, 112106. [CrossRef]
24. Bruno-Alfonso, A.; Latgé, A. Aharonov-Bohm oscillations in a quantum ring: Eccentricity and electric-field effects. *Phys. Rev. B* **2005**, *71*, 125312. [CrossRef]
25. Mielnik, B. Evolution loops. *J. Math. Phys.* **1986**, *27*, 2290–2306. [CrossRef]
26. Delgado, F.; Mielnik, B. Are there Floquet quanta? *Phys. Lett. A* **1998**, *249*, 369–375. [CrossRef]

27. Pinney, E. The nonlinear differential equation $y'' + p(x)y + cy^{-3} = 0$. *Proc. Am. Math. Soc.* **1950**, *1*, 581. [CrossRef]
28. Ermakov, V.P. Second order differential equations. Conditions to complete integrability. *Kiev Univ. Izvestia Ser. III* **1880**, *9*, 125. (In Russian) [CrossRef]
29. Klimov, A.B.; Chumakov, S.M. *A Group-Theoretical Approach to Quantum Optics*; Wiley-VCH: Weinheim, Germany, 2009.
30. Gutiérrez-Vega, J.C.; Rodríguez-Dagnino, R.M.; Meneses-Nava, M.A.; Chávez-Cerda, S. Mathieu functions, a visual approach. *Am. J. Phys.* **2003**, *71*, 233–242. [CrossRef]
31. Haroche, S.; Raimond, J.M. *Exploring the Quantum: Atoms, Cavities, and Photons*; Oxford University Press: Oxford, UK, 2006.
32. Anandan, J. The geometric phase. *Nature* **1992**, *360*, 307–313. [CrossRef]
33. Cohen, E.; Larocque, H.; Bouchard, F.; Nejadsattari, F.; Gefen, Y.; Karimi, E. Geometric phase from Aharonov-Bohm to Pancharatnam-Berry and beyond. *Nat. Rev. Phys.* **2019**, *1*, 437–449. [CrossRef]
34. Zanardi, P.; Rasetti, M. Holonomic quantum computation. *Phys. Lett. A* **1999**, *264*, 94–99. [CrossRef]
35. Jones, J.A.; Vedral, V.; Ekert, A.; Castagnoli, G. Geometric quantum computation using nuclear magnetic resonance. *Nature* **2000**, *403*, 869–871. [CrossRef] [PubMed]
36. Aharonov, Y.; Anandan, J. Phase change during a cyclic quantum evolution. *Phys. Rev. Lett.* **1987**, *58*, 1593–1596. [CrossRef] [PubMed]
37. Menda, I.; Buric, N.; Popovic, D.B.; Prvanovic, S.; Radonjic, M. Geometric Phase for Analytically Solvable Driven Time-Dependent Two-Level Quantum Systems. *Acta Phys. Polonica* **2014**, *126*, 670–672. [CrossRef]
38. Chruściński, D.; Jamiołkowski, A. *Geometric Phases in Classical and Quantum Mechanics*; Birkhäuser: Boston, MA, USA, 2004.

© 2019 by the authors. Licensee MDPI, Basel, Switzerland. This article is an open access article distributed under the terms and conditions of the Creative Commons Attribution (CC BY) license (http://creativecommons.org/licenses/by/4.0/).

Article

Asymmetry of Quantum Correlations Decay in Nonlinear Bosonic System

Anna Kowalewska-Kudłaszyk *,† and Grzegorz Chimczak †

Nonlinear Optics Division, Faculty of Physics, Adam Mickiewicz University, Uniwersytetu Poznańskiego 2, 61-614 Poznań, Poland
* Correspondence: annakow@amu.edu.pl
† These authors contributed equally to this work.

Received: 12 July 2019; Accepted: 6 August 2019; Published: 8 August 2019

Abstract: We study the problem of the influence of one-sided different noisy channels to the quantum correlations decay in a symmetric bosonic system. We concentrate on one type of these correlations—the entanglement. The system under consideration is composed of two nonlinear oscillators coupled by two-boson interactions and externally driven by a continuous coherent field. Our low-dimensional system can be treated as 2-qutrit one. Two different noisy channels (the amplitude and the phase-damping reservoirs) are applied to both of the system's modes. We show that there is a noticeable difference in the quantum entanglement in 2-qubit subspaces of the whole system decrease after swapping the reservoirs between the modes of the considered symmetric system. It appears also that the degree of obtained entanglement depends crucially on the position of the appropriate type of reservoir. The origin of the observed asymmetry is also explained.

Keywords: nonlinear oscillator; quantum entanglement; open system

---

1. Introduction

Obtaining low-dimensional optical systems is still one of the challenges in contemporary quantum optics. Systems evolving among a strictly limited number of quantum states play an important role, e.g., in quantum cryptography [1,2], state teleportation [3,4], and quantum information processing [5].

The methods leading to the state truncation rely mainly on the applications of nonlinear media of various types. In the literature we may find both theoretical and experimental examples of low-dimensional bosonic systems obtained with the help of effective nonlinearities of various orders and strengths [6–9]. One of such methods makes use of quantum effects leading to the phenomena known as photon blockade (or more generally, boson blockade). This is the phenomenon in which the Hilbert space of the considered system is significantly truncated in such a way that only a very limited number of bosonic states is available for the whole system. Such a truncation can be obtained in two different ways, and two different physical mechanisms lead to photon blockades.

In the first mechanism, which is known as *conventional photon blockade*, the origin of truncation lies in obtaining an effective nonlinearity in the Hamiltonian describing the system [10–13]. This nonlinearity is responsible for the non-equidistant eigenstates spectrum of the system. An external driving resonant with one transition is not resonant with others, and thus, the number of allowed states is limited (truncated). For example, the effective nonlinearity of a cavity field–atom interaction leads to the anharmonicity of the Jaynes–Cummings ladder of eigenstates. The presence of single photon in this optical cavity suppresses the appearance of the second one in the system, because a second photon at the same frequency is not resonant with the next transitions in the ladder [10]. This mechanism works properly only if the second photon is detuned from each of those next transitions by amounts that are much larger than the respective linewidths. Therefore, strong nonlinearities, and thus strong light–matter regimes, are necessary in this first mechanism. Large nonlinearities may be problematic from an experimental point of view, but many examples have already been presented in which strong effective nonlinearities are generated, e.g., due to: cavity field–atom interactions in the dispersive limit [10], optomechanical interactions [14], interactions between the atoms in optical lattices [15], nanoresonators [16], and interactions in superconducting circuit-QED systems [17,18].

In the weak light–matter regime, where the effective nonlinearity is weak, the second mechanism (known as *unconventional photon blockade*) is necessary [19–21]. It is a strongly resonant effect and requires a minimum input intensity to operate [22]. The physical origin of this blockade effect is the result of multipath destructive interference, which is responsible for uncoupling of some of the states.

In the considerations presented in this paper, we apply the systems characterized by strong nonlinearities to make use of the physical mechanism leading to the conventional photon blockade. Such systems are also known in the literature as *nonlinear quantum scissors*, due to their properties resulting in truncation of the Hilbert space [23–25]. Due to the presence of strong nonlinearity, which changes the energy spectrum of the considered system, there is only a limited number of resonant states, and finally the effective state truncation is possible.

In many papers the limited number of two-mode states is the starting point of the discussion of entanglement formation. Such possibility was reported for example in papers dealing with nonlinear quantum scissors, which exploited strongly nonlinear media (with $\chi^{(3)}$ much larger than other parameters describing the interactions) to obtain 2-qubit, qutrit-qubit or 2-qutrit systems [26–28]. Various external driving mechanisms were also analyzed: continuous excitation, performed by a laser field [24,26]; intense optical field generating two modes by parametric down-conversion process [27]. Pulsed laser driving with constant or different pulse widths was also analyzed [29]. In all these cases, generation of maximally or almost maximally entangled states were reported. It was also shown that changes in the duration of laser pulse or even random perturbations in pulse duration can improve the ability of the nonlinear system to produce entangled states [29]. It may be of special worth especially for real experimental situations when some perturbations in maintaining the precise and stable timing of pulse laser sources can occur.

In the present paper, a coupled oscillatory system evolving among limited number of quantum states externally driven in both modes by a coherent field is presented. We concentrate on an open system evolution of coupled nonlinear media and apply different (amplitude and phase) noisy channels to both modes of the considered system. We show that we may influence the rate of disentanglement and the values of entanglement created in the qubit subspaces of the whole system by swapping the noisy channels between the modes. Such changes in the environment type, which influence the entangled qubits, may occur when both parts of the entangled pair are sent through different paths and therefore may encounter various local disturbances. The observed asymmetry in the response of the analyzed 2-mode system may give some information on which type of environment the system's parts were exposed to and which of the qubits was affected by a specific type of noise. The influence of the reservoir and the reservoir combinations on the symmetry of quantum correlation were analyzed for example in [30] for some mixed states under the influence of depolarizing channel. For the specified X state, the problem of symmetry in correlations decay was reported for example in [31].

2. The Model

The systems we are dealing with consist of coupled quantum nonlinear media with external excitations, which in general can be of various types. Such systems are well known as the potential sources of maximally or almost maximally entangled states. Various types of coupling and external driving have already been discussed [26–28,32]. In all of those cases it appeared that it was possible to restrict the evolution of the whole coupled system to several bosonic states only—therefore, under the assumptions of weak interactions (as compare with the nonlinearity), truncation of the appropriate wave function was possible. In consequence it has also been shown that such nonlinear systems can be treated as pairs of qubits, or qutrits or qutrit-qubit pairs. It has also been shown that under some assumptions, creation of for example Bell-type [24,26] or W-states [32] was possible.

Such models can be realized either in the quantum optical domain with media characterized by ultra-strong Kerr nonlinearities, or more often in the systems described by effective Hamiltonians including nonlinearities of the Kerr type.

Each of the parts forming the coupled system is characterized by the following nonlinear Hamiltonian in the interaction picture:

$$\hat{H}_{NL} = \frac{\chi_a^{(3)}}{2}\hat{a}^{\dagger 2}\hat{a}^2 + \frac{\chi_b^{(3)}}{2}\hat{b}^{\dagger 2}\hat{b}^2 = \frac{\chi_a^{(3)}}{2}\hat{n}_a(\hat{n}_a - 1) + \frac{\chi_b^{(3)}}{2}\hat{n}_b(\hat{n}_b - 1), \tag{1}$$

where a and b label two modes of the system, $\hat{n}_a$ and $\hat{n}_b$ are the photon number operators in those modes. In general, both parts can have different nonlinear properties: $\chi_a \neq \chi_b$.

In our considerations of the decay asymmetry we will concentrate on the nonlinear exchange between the bosonic modes. In the optical counterparts of such a system that type of interaction is realized by two-photon processes:

$$\hat{H}_{int_{nonl}} = \epsilon(\hat{a}^{\dagger})^2\hat{b}^2 + \epsilon^*(\hat{b}^{\dagger})^2\hat{a}^2, \tag{2}$$

where ϵ describes the strength of the mutual interaction. Apart from the interactions between nonlinear parts, we assume that the system is driven by coherent laser fields with intensities α and β. Therefore, the appropriate Hamiltonian has the following form:

$$\hat{H}_{ext} = \alpha\hat{a}^{\dagger} + \alpha^*\hat{a} + \beta\hat{b}^{\dagger} + \beta^*\hat{b}. \tag{3}$$

Under assumptions of weak internal and external interactions (if compared to Kerr-type nonlinearity), the evolution of the discussed system is closed within a few two-mode states only, despite the continuous excitations. To make use of the nonlinear bosonic exchange between the modes of a coupled oscillatory system we prepare it in the initial excited state $|\Psi(t=0)\rangle = |0\rangle_a|2\rangle_b$, and when symmetrically exciting both modes the adequate wave function takes the following form:

$$|\Psi(t)\rangle_{cut_{nonl}} = c_{02}(t)|0\rangle_a|2\rangle_b + c_{12}(t)|1\rangle_a|2\rangle_b + c_{20}(t)|2\rangle_a|0\rangle_b + c_{21}(t)|2\rangle_a|1\rangle_b, \tag{4}$$

and forms a 2-qutrit system, as in both modes only states $|0\rangle$, $|1\rangle$ and $|2\rangle$ are populated. The appropriate fidelity for obtaining the wave function (4) is plotted in Figure 1. We calculate fidelities between the two wave functions of the system: the truncated wave function for the specified subspace and the whole numerically obtained wave function in an extended basis. We can easily see that our nonlinear coupled system evolves only between the states belonging to two subspaces: $\{|0\rangle_a|0\rangle_b; |0\rangle_a|2\rangle_b; |2\rangle_a|0\rangle_b; |2\rangle_a|2\rangle_b;\}$ (solid line) and $\{|1\rangle_a|1\rangle_b; |1\rangle_a|2\rangle_b; |2\rangle_a|1\rangle_b; |2\rangle_a|2\rangle_b;\}$ (dashed line). Therefore, the influence of the states other than (4) is negligible and we can treat the state truncation as fully justified.

The analysis of formation of maximally entangled states within the system, given in [25,26], enables us to claim that the coupled oscillators form the following entangled states with probabilities equal to unity ($|B\rangle_{1(2)}$) or slightly less than unity ($|B\rangle_3$ for which fidelity ≈ 0.98). As also the mode b is externally driven by a coherent field, additionally the state $|2\rangle_a|1\rangle_b$ is included in the evolution. Consequently, also states $|B\rangle_{3(4)}$ can be obtained, but with lower probabilities. These maximally entangled states are defined by:

$$|B\rangle_{1(2)} = \frac{1}{\sqrt{2}}\left(|2\rangle_a|0\rangle_b \pm i|0\rangle_a|2\rangle_b\right), \qquad (5)$$

$$|B\rangle_{3(4)} = \frac{1}{\sqrt{2}}\left(|2\rangle_a|1\rangle_b \pm i|1\rangle_a|2\rangle_b\right). \qquad (6)$$

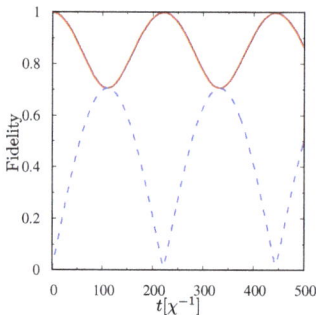

Figure 1. Fidelity between the numerically obtained wave function describing the whole two-mode system in an extended basis and the truncated wave function for the subspace $\{|0\rangle_a|0\rangle_b; |0\rangle_a|2\rangle_b; |2\rangle_a|0\rangle_b; |2\rangle_a|2\rangle_b;\}$ — solid line, and for the subspace $\{|1\rangle_a|1\rangle_b; |1\rangle_a|2\rangle_b; |2\rangle_a|1\rangle_b; |2\rangle_a|2\rangle_b;\}$ — dashed line. The initial state of the system is a two-photon state $|0\rangle_a|2\rangle_b$, $\gamma = 0$; $\epsilon = \alpha = \chi/100$.

3. Entanglement Decay under Dissipation Channel Combinations

The analysis of the influence of various dissipation channels on the system's dynamics can be described by standard techniques when considering master equation for the density matrix of the system. For our purpose we will assume two-sided different noisy channels. We will focus on the amplitude and phase channels applied to both of the system's modes. In particular, we will concentrate on the problem of the decay of the created entangled states for the symmetrically driven system. The amplitude damping in Born and Markov approximations for a specified mode can be obtained by adding an appropriate Liouvillian to the master equation:

$$\frac{d}{dt}\hat{\rho} = -i(\hat{H}\hat{\rho} - \hat{\rho}\hat{H}) + \sum_{k=a,b}\gamma_k\left[\hat{k}\hat{\rho}\hat{k}^\dagger - \frac{1}{2}\left(\hat{\rho}\hat{k}^\dagger\hat{k} + \hat{k}^\dagger\hat{k}\hat{\rho}\right)\right]. \qquad (7)$$

For the phase-damping channel, the adequate master equation has the following form:

$$\frac{d}{dt}\hat{\rho} = -i(\hat{H}\hat{\rho} - \hat{\rho}\hat{H}) + \frac{1}{2}\sum_{k=a,b}\gamma_k\left[2\hat{k}^\dagger\hat{k}\hat{\rho}\hat{k}^\dagger\hat{k} - \left(\hat{k}^\dagger\hat{k}\right)^2\hat{\rho} - \hat{\rho}\left(\hat{k}^\dagger\hat{k}\right)^2\right], \qquad (8)$$

where $\hat{k}$ denotes the specified mode of the system, and we additionally assume zero temperature reservoirs. The Hamiltonian $\hat{H}$ consists of the nonlinear part (1), the Hamiltonian (2) describing the interactions within the oscillatory system, and the Hamiltonian (3) with excitations performed in both modes. The whole coupled system is fully symmetric, as depicted in Figure 2, and for our analysis we will assume switching the location of the reservoir with specified type of noise between the qubits which we can create the entangled states and look for the symmetry in the quantum correlations dynamics.

We will focus on one type of quantum correlations, namely the entanglement. For the analysis of entanglement, we can apply negativity [33,34], which is a measure of entanglement degree in 2-qubit and qubit-qutrit systems

$$N(\rho) = \frac{1}{2}\sum_i |\lambda_i| - \lambda_i, \qquad (9)$$

where λ_i is the i-th eigenvalue of a matrix ρ^{T_k}. This matrix is obtained by performing partial transpose of the matrix ρ of the whole quantum system with respect to one of the subsystems, a or b. Using that measure we can identify the maximally entangled states (in the systems of 2 qubits and qubit-qutrit) as those for which the negativity equal to unity is obtained. $N(\rho) = 0$ holds for separable states. In further considerations we will analyze the process of entanglement creation within the 2-qubit subsystem of the whole system. Therefore we restrict ourselves to the subspace spanned by the following two-mode states: $\{|0\rangle_a|0\rangle_b; |0\rangle_a|2\rangle_b; |2\rangle_a|0\rangle_b; |2\rangle_a|2\rangle_b;\}$; N_{0220} is the abbreviation for the negativity defined according to (9), with the matrix ρ describing this 2-qubit subsystem. The second 2-qubit subsystem is related to the states: $\{|1\rangle_a|1\rangle_b; |1\rangle_a|2\rangle_b; |2\rangle_a|1\rangle_b; |2\rangle_a|2\rangle_b;\}$ and the appropriate negativity N_{1221}.

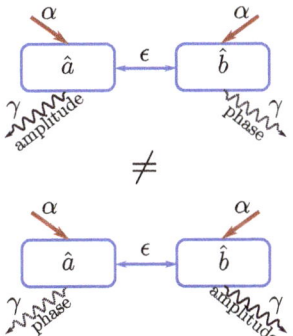

Figure 2. Visualization of different location of noise in the model of the symmetric coupled oscillatory system.

It is well known that the amplitude reservoir describes the process of energy dissipation induced by the environment. Therefore, the decrease in the states' populations leads consequently to the decay of quantum correlations such as for example quantum entanglement. Phase-damping reservoir, on the other hand, describes the process of a random change in time of relative phases of the superposed states. In that process the number of photons remains unchanged and the energy is preserved. Nevertheless, the dephasing process is also responsible for quantum correlations destruction, but without changes of the total energy of the system. When looking at the density matrix—the phase-damping channel affects the off-diagonal elements only, while the amplitude damping channel influences the diagonal as well as the off-diagonal elements.

In [35] the influence of symmetrically applied amplitude and phase-damping channels was analyzed and it was shown that the dephasing channel, although it leads to the decrease of entanglement (measured by the negativity) but the rate of such a process is much slower than for the system exposed to the amplitude damping environment. Also, the entanglement sudden death, observed when energy dissipation is allowed, was not possible due to the dephasing process only. Another interesting feature of the analyzed qutrit-qubit system was also noticed. Specifically, the state $|1\rangle_a|2\rangle_b$ appeared to be helpful in restoring the degree of entanglement when the system is exposed to the amplitude damping process. It is possible because of constant external driving and formation of the entangled state $|B\rangle = \frac{1}{\sqrt{2}}\left(|2\rangle_a|0\rangle_b + i|1\rangle_a|2\rangle_b\right)$, which involves $|2\rangle_a|0\rangle_b$. Therefore, the negativity can be periodically increased due to the presence of correlations, which depend on the external coherent field of the amplitude α. Correlations can therefore flow out of the considered 2-qubit subspace and return, due to the interactions with the additional two-mode state. Actually, our coupled nonlinear system is a higher-dimensional one. When only unitary evolution is assumed, the wave function can be expressed by (4), therefore we must consider 2-qutrits instead of just a qutrit-qubit system. Apart from the state $|1\rangle_a|2\rangle_b$ there is also nonzero probability of obtaining the state $|2\rangle_a|1\rangle_b$, and other entangled states $|B\rangle_{3(4)}$ are also possible to obtain.

The state $|2\rangle_a|2\rangle_b$ is a common state for both of the 2-qubit subspaces, therefore it is evident that they are not independent and correlations obtained in one of them may influence the correlations observed in the other one.

We assume that the whole system is fully symmetrical — nonlinear media are of the same type, $\chi_a = \chi_b$, they are symmetrically driven with the same strengths $\alpha = \beta$ (see Figure 2). We apply different damping channels to both modes a and b and look for any differences under swapping the channels, which affect both parts of the whole considered system.

Still we concentrate mainly on the entanglement created only between the states $\{|0\rangle_a|0\rangle_b; |0\rangle_a|2\rangle_b; |2\rangle_a|0\rangle_b; |2\rangle_a|2\rangle_b;\}$ (N_{0220}) and between the states $\{|1\rangle_a|1\rangle_b; |1\rangle_a|2\rangle_b; |2\rangle_a|1\rangle_b; |2\rangle_a|2\rangle_b;\}$ (N_{1221}). We are going to ask ourselves the question, whether it is possible in a real physical system composed of these states to see any asymmetrical behavior of negativity decay by switching the damping channels between the modes a and b. The appropriate figures (Figure 3) presenting the time dependence of negativities calculated for 2-qubit subspace for various relations between the mutual (ε) and the external (α) interactions, are presented. First of all, it can be easily seen that there is an evident difference between the values of negativities obtained and the rate of decay under the symmetric channel swapping. Therefore, a fully symmetrical quantum system can exhibit a clear asymmetry in evolution of quantum correlations when applying and swapping different types of reservoirs.

As seen in Figure 3, for all the considered cases larger values of entanglement characterized by N_{0220} and N_{1221}, are possible, when the mode a is exposed to the amplitude damping channel and the mode b decays to the phase reservoir (Figure 3a,c), than when the reservoirs are swapped between the modes (Figure 3b,d). There is also a noticeable difference in the time span for which the system can be considered to be an entangled one. Again, longer times for disentangling of the qubits are necessary if the mode a decays to the reservoir with amplitude damping and the mode b to the phase than for the case with the mode a decaying to the dephasing channel and the mode b to the amplitude channel. Therefore, there is a certain asymmetry in the response of the coupled system to the exchange of the damping channels between the parts of the whole system. We can also see that the presence of correlations is connected mainly with the subspace $\{|0\rangle_a|0\rangle_b; |0\rangle_a|2\rangle_b; |2\rangle_a|0\rangle_b; |2\rangle_a|2\rangle_b;\}$. In all the considered cases (in various damping channels configurations) the values of negativities N_{0220} are noticeably higher than appropriate negativities N_{1221} for the second considered subspace. Additionally, the decrease in the value of N_{0220} is obviously connected with the increase of N_{1221}. Thus, there is some kind of flow of entanglement between the considered subspaces of the whole system.

Next we will address the problem of the specific entangled states (5)–(6) formation to decide, which of them are responsible for nonzero negativities and which influence the observed asymmetry in $N_{0220}(t)$ and $N_{1221}(t)$ dependencies after exchanging the reservoirs between the modes a and b.

It is evident that all the states (5)–(6) should be possible to observe during the time evolution of the system. States $|B\rangle_{1(2)}$ are created due to the 2-photon interactions, states $|B\rangle_{3(4)}$ are obtained as a consequence of additional external excitation, which is responsible for populating the states $|1\rangle_a|2\rangle_b$ and $|2\rangle_a|1\rangle_b$.

In Figure 4 the evolutions of appropriate entangled states are plotted. When the mode a is decaying into the amplitude damping channel, the states $|B\rangle_{1(2)}$ oscillate with almost the same frequencies, simultaneously obtaining maxima and minima. When the mode a is exposed to the dephasing process, the evolution of $|B\rangle_2$ state is slightly perturbed by another frequency. From the behavior of entangled states presented, it comes out that a complete decay of negativity is related to the higher probability of obtaining the states $|B\rangle_{3(4)}$ —in such a case the correlations initially created in the subspace $\{|0\rangle_a|0\rangle_b; |0\rangle_a|2\rangle_b; |2\rangle_a|0\rangle_b; |2\rangle_a|2\rangle_b;\}$ are transferred out of the considered subspace. Due to the presence of one common state belonging to both subspaces of the whole system, correlations can be significantly transferred to the subspace $\{|1\rangle_a|1\rangle_b; |1\rangle_a|2\rangle_b; |2\rangle_a|1\rangle_b; |2\rangle_a|2\rangle_b;\}$. It happens in shorter time for the situation when the mode a is exposed to the dephasing process, as it is easier to populate the state $|1\rangle_a|2\rangle_b$ by an external field, while the mode a does not lose its population. The presence of the state $|1\rangle_a|2\rangle_b$ is therefore crucial in obtaining the asymmetry in correlations, measured by the negativity decay introduced by swapping the reservoirs between the qubits forming the entangled states. As seen from Figure 4, the evolution of the created entangled states strongly depends on the type of reservoir applied to each of the modes. Such a difference is obviously seen in the values of negativities for the qubit subspaces and results in the asymmetry in correlations decay.

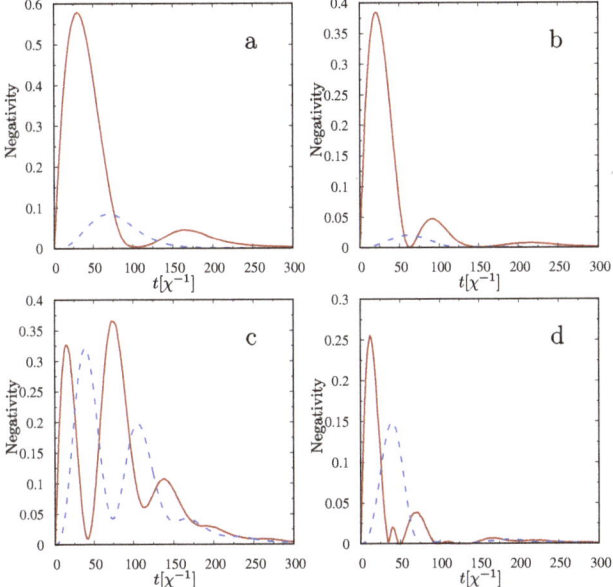

Figure 3. Negativities for 2-qubit subspaces—N_{0220} (solid line) and N_{1221} (dashed line) versus time scaled in $1/\chi$ units for combinations of damping channels. In (**a**) and (**c**) the mode a is exposed to the amplitude damping channel and mode b into the phase-damping channel. In (**b**) and (**d**) the channels are swapped. The initial state of the system is a two-photon one $|0\rangle_a|2\rangle_b$, $\gamma = \epsilon = \alpha = \chi/100$ for (**a**) and (**b**); $\gamma = \epsilon = \chi/100$ and $\alpha = \chi/25$ for (**c**) and (**d**).

It appears that the state $|1\rangle_a|2\rangle_b$ plays a crucial role in formation and behavior of the entanglement in the coupled system. Its presence influences significantly the time of disentanglement and indirectly the values of N_{0220} — higher probabilities of $|B\rangle_{3(4)}$ decrease the probabilities of $|B\rangle_{1(2)}$. When the mode a of the state $|1\rangle_a|2\rangle_b$ decays by the amplitude channel to the state $|0\rangle_a|2\rangle_b$, it increases the probability of generating the states $|B\rangle_{1(2)}$ and results in higher values of N_{0220}. On the other hand, the mode a is externally pumped by the linear interaction and again the state $|1\rangle_a|2\rangle_b$ can be formed.

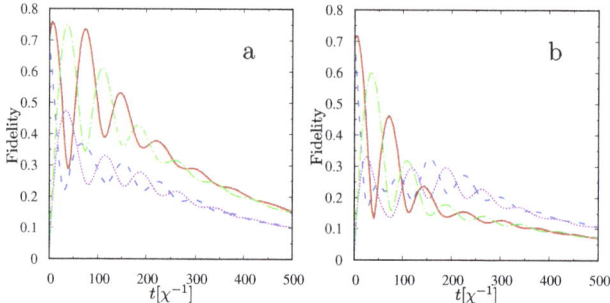

Figure 4. Fidelities for obtaining the entangled states for $\gamma = \epsilon = \chi/100$, $\alpha = \chi/25$ and the initial state $|0\rangle_a |2\rangle_b$. In (**a**) the mode a is exposed to the amplitude damping channel and the mode b to the phase-damping channel. In (**b**) the channels are swapped. $|B\rangle_1$—solid line; $|B\rangle_2$—dashed line; $|B\rangle_3$—dashed-dotted line and $|B\rangle_4$—dotted line.

4. Conclusions

We discussed a coupled nonlinear system whose parts are driven by coherent laser fields. We additionally assumed that interactions between the nonlinear media are performed via two-boson exchange. Such systems, when appropriate conditions describing relations between the strengths of the interactions and the nonlinearity are fulfilled, evolve only between a limited number of states despite the constant energy supply. We have shown that the model we are dealing with can be treated as a 2-qutrit system and the unitary evolution can be described by a truncated wave function. As the nonlinearity describing the interacting media must be large (as compared to other interactions), in that sense the possibility of the Hilbert space truncation may be regarded as the conventional photon blockade. We have discussed the problem of the entanglement decay under the influence of different noisy channels. Both parts of the considered coupler (the mode a and the mode b) were exposed to the local amplitude or phase-damping environments. Our goal was to show that despite the symmetry of our system, it is possible to differentiate the decay rates and the values of the entanglement obtained within 2-qubits subspaces by appropriately choosing the type of system-environment interactions. For that purpose, we have analyzed the time evolution of negativities after first applying and then swapping different noisy channels to the modes a and b. We have identified the asymmetry in the response of the system to the interactions with the reservoirs. There is a noticeable difference in the rate at which entanglement decreases, due to the location of the noise between the qubits.

Author Contributions: Conceptualization and Formal analysis, A.K.-K.; Software, Validation and Investigation, A.K.-K. and G.C.; Visualization, G.C.; Writing—original draft, A.K.-K.; Writing—review & editing, G.C.

Funding: This research received no external funding

Conflicts of Interest: The authors declare no conflict of interest.

References

1. Gisin, N.; Ribordy, G.; Tittel, W.; Zbinden, H. Quantum cryptography. *Rev. Mod. Phys.* **2002**, *74*, 145–195. [CrossRef]
2. Bartkiewicz, K.; Černoch, A.; Chimczak, G.; Lemr, K.; Miranowicz, A.; Nori, F. Experimental quantum forgery of quantum optical money. *npj Quantum Inf.* **2017**, *3*, 7. [CrossRef]
3. Özdemir, Ş.K.; Bartkiewicz, K.; Liu, Y.X.; Miranowicz, A. Teleportation of qubit states through dissipative channels: Conditions for surpassing the no-cloning limit. *Phys. Rev. A* **2007**, *76*, 042325. [CrossRef]

4. Chimczak, G.; Tanaś, R. High-fidelity atomic-state teleportation protocol with non-maximally-entangled states. *Phys. Rev. A* **2009**, *79*, 042311. [CrossRef]
5. Chimczak, G. Efficient generation of distant atom entanglement via cavity decay. *Phys. Rev. A* **2005**, *71*, 052305. [CrossRef]
6. D'Ariano, G.M.; Maccone, L.; Paris, M.; Sacchi, M. Optical Fock-state synthesizer. *Phys. Rev. A* **2000**, *61*, 053817. [CrossRef]
7. Didier, N.; Pugnetti, S.; Blanter, Y.M.; Fazio, R. Detecting phonon blockade with photons. *Phys. Rev. B* **2011**, *84*, 054503. [CrossRef]
8. Wang, H.; Gu, X.; Liu, Y.X.; Miranowicz, A.; Nori, F. Tunable photon blockade in a hybrid system consisting of an optomechanical device coupled to a two-level system. *Phys. Rev. A* **2015**, *92*, 033806. [CrossRef]
9. Peřinová, V.; Lukš, A.; Křapelka, J. Dynamics of nonclassical properties of two- and four-mode Bose-Einstein condensates. *J. Phys. B At. Mol. Opt. Phys.* **2013**, *46*, 195301. [CrossRef]
10. Birnbaum, K.M.; Boca, A.; Miller, R.; Boozer, A.D.; Northup, T.E.; Kimble, H.J. Photon blockade in an optical cavity with one trapped atom. *Nature* **2005**, *436*, 87–90. [CrossRef]
11. Faraon, A.; Fushman, I.; Englund, D.; Stoltz, N.; Petroff, P.; Vučković, J. Coherent generation of non-classical light on a chip via photon-induced tunnelling and blockade. *Nat. Phys.* **2008**, *4*, 859–863. [CrossRef]
12. Miranowicz, A.; Bajer, J.; Paprzycka, M.; Liu, Y.X.; Zagoskin, A.M.; Nori, F. State-dependent photon blockade via quantum-reservoir engineering. *Phys. Rev. A* **2014**, *90*, 033831. [CrossRef]
13. Zhu, C.J.; Yang, Y.P.; Agarwal, G.S. Collective multiphoton blockade in cavity quantum electrodynamics. *Phys. Rev. A* **2017**, *95*, 063842. [CrossRef]
14. Bose, S.; Jacobs, K.; Knight, P.L. Preparation of nonclassical states in cavities with a moving mirror. *Phys. Rev. A* **1997**, *56*, 4175–4186. [CrossRef]
15. Islam, R.; Ma, R.; Preiss, P.M.; Tai, M.E.; Lukin, A.; Rispoli, M.; Greiner, M. Measuring entanglement entropy in a quantum many-body system. *Nature* **2015**, *528*, 77–83. [CrossRef] [PubMed]
16. Liu, Y.X.; Miranowicz, A.; Gao, Y.B.; Bajer, J.; Sun, C.P.; Nori, F. Qubit-induced phonon blockade as a signature of quantum behavior in nanomechanical resonators. *Phys. Rev. A* **2010**, *82*, 032101. [CrossRef]
17. Hoffman, A.J.; Srinivasan, S.J.; Schmidt, S.; Spietz, L.; Aumentado, J.; Türeci, H.E.; Houck, A.A. Dispersive Photon Blockade in a Superconducting Circuit. *Phys. Rev. Lett.* **2011**, *107*, 053602. [CrossRef]
18. Liu, Y.; Xu, X.; Miranowicz, A.; Nori, F. From blockade to transparency: Controllable photon transmission through a circuit-QED system. *Phys. Rev. A* **2014**, *89*, 043818. [CrossRef]
19. Liew, T.C.H.; Savona, V. Single Photons from Coupled Quantum Modes. *Phys. Rev. Lett.* **2010**, *104*, 183601. [CrossRef]
20. Bamba, M.; Imamoğlu, A.; Carusotto, I.; Ciuti, C. Origin of strong photon antibunching in weakly nonlinear photonic molecules. *Phys. Rev. A* **2011**, *83*, 021802. [CrossRef]
21. Hamsen, C.; Tolazzi, K.N.; Wilk, T.; Rempe, G. Two-Photon Blockade in an Atom-Driven Cavity QED System. *Phys. Rev. Lett.* **2017**, *118*, 133604. [CrossRef] [PubMed]
22. Flayac, H.; Gerace, D.; Savona, V. An all-silicon single-photon source by unconventional photon blockade. *Sci. Rep.* **2015**, *5*, 11223. [CrossRef] [PubMed]
23. Leoński, W.; Tanaś, R. Possibility of producing the one-photon state in a kicked cavity with a nonlinear Kerr medium. *Phys. Rev. A* **1994**, *49*, R20–R23. [CrossRef] [PubMed]
24. Leoński, W.; Miranowicz, A. Kerr nonlinear coupler and entanglement. *J. Opt. B* **2004**, *6*, S37–S42. [CrossRef]
25. Leoński, W.; Kowalewska-Kudłaszyk, A. Quantum Scissors–Finite-Dimensional States Engineering. In *Progress in Optics*; Wolf, E., Ed.; Elsevier: Amsterdam, The Netherlands, 2011; Volume 56, pp. 131–185. [CrossRef]
26. Kowalewska-Kudłaszyk, A.; Leoński, W. Finite-dimensional states and entanglement generation for a nonlinear coupler. *Phys. Rev. A* **2006**, *73*, 042318. [CrossRef]
27. Kowalewska-Kudłaszyk, A.; Leoński, W.; Peřina, J., Jr. Photon-number entangled states generated in Kerr media with optical parametric pumping. *Phys. Rev. A* **2011**, *83*, 052326. [CrossRef]
28. Kowalewska-Kudłaszyk, A.; Leoński, W. Nonlinear coupler operating on Werner-like states; entanglement creation, its enhancement, and preservation. *J. Opt. Soc. Am. B* **2014**, *31*, 1290–1297. [CrossRef]

29. Kalaga, J.; Kowalewska-Kudłaszyk, A.; Jarosik, M.; Szczęśniak, R.; Leoński, W. Enhancement of the entanglement generation via randomly perturbed series of external pulses in a nonlinear Bose–Hubbard dimer. *Nonlinear Dyn.* **2019**, 1–15. [CrossRef]
30. Życzkowski, K.; Horodecki, P.; Horodecki, M.; Horodecki, R. Dynamics of quantum entanglement. *Phys. Rev. A* **2001**, *65*, 012101. [CrossRef]
31. Lyyra, H.; Karpat, G.; Li, C.; Guo, G.; Piilo, J.; Maniscalco, S. Symmetry in the open-system dynamics of quantum correlations. *Sci. Rep.* **2017**, *7*, 8367. [CrossRef]
32. Said, R.S.; Wahiddin, M.R.B.; Umarov, B.A. Generation of three-qubit entangled W state by nonlinear optical state truncation. *J. Phys. B At. Mol. Opt. Phys.* **2006**, *39*, 1269–1274. [CrossRef]
33. Peres, A. Separability Criterion for Density Matrices. *Phys. Rev. Lett.* **1996**, *77*, 1413–1415. [CrossRef] [PubMed]
34. Horodecki, M.; Horodecki, P.; Horodecki, R. Separability of mixed states: Necessary and sufficient conditions. *Phys. Lett. A* **1996**, *223*, 1–8. [CrossRef]
35. Kowalewska-Kudłaszyk, A. Dephasing in nonlinear quantum scissors systems. *Opt. Commun.* **2012**, *285*, 5543–5548. [CrossRef]

© 2019 by the authors. Licensee MDPI, Basel, Switzerland. This article is an open access article distributed under the terms and conditions of the Creative Commons Attribution (CC BY) license (http://creativecommons.org/licenses/by/4.0/).

Article

The Symmetry of Pairing and the Electromagnetic Properties of a Superconductor with a Four-Fermion Attraction at Zero Temperature

Przemyslaw Tarasewicz

Faculty of Pharmacy, Collegium Medicum in Bydgoszcz Nicolaus Copernicus University in Toruń, ul. Jagiellońska 13, 85-067 Bydgoszcz, Poland; tarasek1@cm.umk.pl; Tel.: +48-52-585-3428; Fax: +48-52-585-3308

Received: 20 September 2019; Accepted: 22 October 2019; Published: 2 November 2019

Abstract: Properties of a fermion system at zero temperature are investigated. The physical system is described by a Hamiltonian containing the BCS interaction and an attractive four-fermion interaction. The four-fermion potential is caused by attractions between Cooper pairs mediated by the phonon field. In this paper, the BCS interaction is assumed to be negligible and the four-fermion potential is the only one that acts in the system. The effect of the pairing symmetry used in the four-fermion potential on some zero-temperature properties is studied. This especially concerns the electromagnetic response of the system to an external magnetic field. It turns out that, in this instance, there are serious differences between the conventional BCS system and the one investigated in this paper.

Keywords: superconductivity; four-fermion attraction; Meissner effect; fermion quartets

PACS: 74.20.-z; 74.20.Fg

---

1. Introduction

Recent decades have been witnessing many new phenomena discovered in solid-state physics (e.g., high temperature superconductivity, heavy fermion superconductors and unusual properties of ^3He). These discoveries are intriguing and lead to many new questions on the nature of superconductivity and superfluidity. The problem of the internal symmetry of gap parameters stands for one such an example. Furthermore, we still do not know how many particles are engaged in constituting the fundamental clusters responsible for the occurrence of new phases. Is it possible that only two-particle entities are relevant to these systems (e.g., Cooper pairs) or may three or four-particle structures also be taken into account? The last question was addressed in [1]. This issue becomes more and more interesting due to some recent discoveries and suggestions. To give some examples, let us start with a work by Schneider and Keller [2] who measured the relationship between the critical temperature and zero temperature condensate density for some cuprates and Chevrel-phases superconductors. They noticed that the experimental data for $YBa_2Cu_3O_{6.602}$, for example, ca be associated with the behavior of a dilute Bose gas. As a consequence, they put the suggestion that Bose condensation of weakly interacting fermion pairs could be a possible mechanism of transition from the normal to the superconducting state. Moreover, a subsequent discovery of Bunkov et al. [3] points to the presence of fermion quartets in ^3He. They investigated the problem of the influence of spatial disorder on the order parameter in superfluid ^3He. They followed Volovik [4] and suggested that unusual spectra of ^3He in an aerogel could be explained by a process in which impurities tend to destroy the anisotropic correlations of the order parameter. Especially interesting is that the correlation of higher symmetry can survive (e.g., four-particle correlations). Two papers [5,6] reported on a discovery of the half-$h/2e$ magnetic flux quanta coexisting together with the usual ones in SQUIDs fabricated of bicrystalline $YBa_2Cu_3O_{7-\delta}$ films. As is widely known, such a circumstance corresponds to the presence of fermion quartets in a physical system. This in turn leads to taking the interplay between Cooper pairs and quartets into consideration. It should be mentioned that there were already some attempts for introducing the four-fermion interactions in the field of nuclear physics [7].

The problem of symmetry of pairing in superconductors has a long history. In classical papers [8,9], this topic was established and studied to a large extent. As is known in the classical BCS theory, the BCS potential binds electrons into Cooper pairs in which pairing has the s-wave symmetry, i.e., the energetic gap that opens at the Fermi level is isotropic in the momentum space. This kind of pairing symmetry is displayed mostly by metallic superconductors. However, there have appeared novel superconductors, e.g., copper oxides and heavy-fermion superconductors. In many of these materials, the order parameter turned out to be anisotropic in the momentum space, i.e., it vanished along some directions. In cuprates $d_{x^2-y^2}$-wave pairing occurs [10] whereas the mixed pairing symmetry $s + id$ is frequently present in heavy-fermion systems [11]. In this paper, we deal with three cases: the pure s-wave pairing, the pure $d_{x^2-y^2}$-wave pairing and the mixed $s + id_{x^2-y^2}$-wave pairing symmetries. We also consider two kinds of electronic bands, i.e., the band with the rectangular density of states (DOS) with the parabolic dispersion relation (the two-dimensional system) and the band with the cosine dispersion relation (the one- and two-dimensional systems). We calculate the order parameter, the ground state energy per lattice site and the electromagnetic kernel as the quantity describing the response of the system to an external magnetic field.

2. The Model

In this paper, some ground state properties of a fermion system, especially in the case of the half-filled band case, are investigated. The Hamiltonian of the system reads

$$H = H_{BCS} + V_4,\qquad(1)$$

where

$$H_{BCS} = \sum_{k\sigma} \xi_k a^*_{k\sigma} a_{k\sigma} - N^{-1} \sum_{k,k'} G_{kk'} a^*_{k+} a^*_{-k-} a_{-k'-} a_{k'+}.$$

The reduced four-fermion interaction V_4 has a form

$$V_4 = -N^{-1} \sum_{k,k'} g_{kk'} b^*_k b^*_{-k} b_{-k'} b_{k'},\qquad(2)$$

where

$$b_k = a_{k+} a_{k-}$$

and operators $a_{k\sigma}$, $a^*_{k\sigma}$ denote the annihilation and creation operators, respectively. N is the number of lattice sites and ξ_k denotes the one-electron energy counted with respect to the chemical potential μ. The coupling functions $G_{kk'}$ and $g_{kk'}$ are assumed to be nonzero in the whole band as it should be in many novel superconductors that are narrow-band systems. As we remember in classical superconductors they are restricted to a narrow shell around the Fermi level. Moreover, it is assumed that both of potentials are attractive so the system is a mixture of fermion pairs with opposite momenta and spins as well as quartets made of these pairs.

In [12], we derived the Hamiltonian (1) by making use of a Fröhlich type transformation of the electron-phonon Hamiltonian. Moreover, we were able to find the sign and the approximated form of coupling constant of the four-fermion interaction.

$$g_{k_F k_F} \approx \frac{D^6_{k_F}}{\hbar^5 \omega^5_{k_F}},$$

where D_{k_F} and $\hbar\omega_{k_F}$ are the electron-phonon coupling constant and the phonon energy at Fermi level, respectively. The relationship between this coupling constant and BCS coupling constant was estimated to be

$$g_{k_F k_F} = \frac{G^3_{k_F k_F}}{\hbar^2 \omega^2_{k_F}}$$

This relationship points to the weak character of the four-fermion interaction in comparison to familiar Cooper pairing. In classical superconductors, quartets cannot be observed due to weak BCS coupling and strong Coulomb repulsion between electrons. However, in materials with strong coupling between electrons and phonons, it seems to be quite reasonable to expect that this phenomenon could be visible and recognizable especially from surveys of magnetic flux quanta. It is widely accepted that if half-$h/2e$ magnetic flux quanta appear among usual ones it points to the existence of quartets in the investigated system. As was established in Introduction these exotic half-$h/2e$ magnetic flux quanta are present in some novel superconductors. These materials possess narrow conduction bands and the antiadiabatic limit ($\frac{\hbar\omega}{t} \gg 1$), where $\frac{\hbar\omega}{t}$ is the energy of an optic phonon while t is the hopping parameter. In this case the starting Hamiltonian could be as follows

$$H = H_0 + H_I,$$

$$H_0 = \sum_{i \neq j, \sigma} t_{ij} c^*_{i\sigma} c_{j\sigma} + (E - \mu) \sum_i n_i + \sum_{i,j} U_{ij} n_i n_j$$

$$H_I = \sum_{i,j} g^s_{ij} n_i (b^*_j + b_j) + \sum_{i,j} g^p_{ij} n_i n_j (b^*_i + b_i),$$

where $c^*_{i\sigma}$ and $c_{i\sigma}$ represent creation and annihilation operators of a local electron from the narrow band. $\sigma = \pm$ denotes the spin of electrons and i and j refer to lattice sites. $n_i = n_{i+} + n_{i-}$ is the number operator. b^*_i and b_i are creation and annihilation operators for a local phonon residing on the i-th site. t_{ij} denotes the hopping integral for electrons in the band. E refer to the site energy of electrons. U_{ij} is the inter-site Coulomb interaction ($i \neq j$) and the local one ($i = j$). g^s_{ij} and g^p_{ij} are the coupling constants of the inter-site phonon–electron potential and the inter-site phonon-electron-electron potential ($i \neq j$). For $i = j$ one gets the local interactions of electrons with phonons. The second term in H_I is based on the argument given by Hirsch et al. [13] that phonons can affect an electron pair sitting on the same lattice site. This is associated with the fact that such a pair affects the shape of an orbital that expands in the space. We extend this argument that two interacting electrons residing on two sites being very close to one another can be influenced by phonons. The term H_I can be removed via making use of the following generalized Lang–Firsov type transformation U^S with the generator

$$S = \sum_{i,j} \left(\frac{g^s_{ij}}{\hbar\omega} n_i + \frac{g^p_{ij}}{\hbar\omega} n_i n_j \right) \left(b^*_j - b_j \right)$$

As a consequence there appear some new purely electronic nonlocal terms, namely, three- and four-electron interactions. We perceive one more possibility to derive a Hamiltonian containing the four-fermion interaction. If one considers the so-called Penson–Kolb–Hubbard model which is described by the following Hamiltonian

$$H_{pkh} = (E - \mu) \sum_{i\sigma} n_{i\sigma} + U \sum_i n_{i+} n_{i-} + W \sum_i n_i n_{i+1} +$$

$$- t^p \sum_i (c^*_{i+} c^*_{i-} c_{i+1-} c_{i+1+} + c^*_{i+1+} c^*_{i+1-} c_{i-} c_{i+}) - t^s \sum_{i\sigma} (c^*_{i\sigma} c_{i+1\sigma} + c^*_{i+1\sigma} c_{i\sigma}),$$

where U and W are coupling constants for intra-site and inter-site Coulomb interactions. $-t^s$ and $-t^p$ are single-electron hopping and pair-hopping integrals, respectively. We admit different signs of U

and W. If the pair-hopping term is weak in comparison to the other potentials, then one can treat it as a perturbation and find the expansion in powers of t^p. We suspect that the four-electron term would appear in the second order of this expansion.

One needs to mention that the problem of the four-fermion interaction in superconductors was investigated in a series of papers by Szczęśniak et al. [14–16]. In some of them, the Eliashberg formalism was applied in order to deal with cuprates as systems with strong coupling of electrons to phonons.

In this paper, we neglect the BCS interaction. This could seem to be strange to do so, however, in [17], it was discovered that for the sufficiently strong four-fermion attraction, the BCS interaction gets turned off. The system in the superconducting state behaves like it would be a condensate of fermion quartets only. The Cooper pairs disappear because they are forced to combine and form fermion quartets. Of course, we are aware of the fact that the scenario based on the four-fermion interaction as the only potential in the system cannot be valid for high-temperature superconductors. If it were the case, then the half-$h/2e$ magnetic flux quanta would be dominating over the conventional ones. So far, this situation has not occurred. Moreover, the four-fermion interaction leads to the first order phase transition whereas novel materials undergo rather the second one. The critical temperatures are lower than those resulting from the BCS theory with the same value of the coupling function [18,19]. This suggests the weaker nature of the four-fermion attraction in comparison to the BCS one. The best solution of this problem is to consider the system with both of potentials and more realistic dispersion relations as it is made in this paper. In [18,19], we make use of the parabolic dispersion relation with the density of states for free electrons that does not seem to be adequate for novel superconductors. Therefore this work describes the limiting case only. The difficulties significantly grow in the case of presence of two attractive potentials in the system.

3. The Mean Field Approximation

In this section, we would like to show the structure of the spectrum of the extended BCS model represented by the Hamiltonian (1). In that paper, we try to address mainly the problem of electrodynamics of the Hamiltonian H defined in the Introduction. The mean field method is used in order to obtain the ground state energy and two-order parameters for Cooper's pairs and quartets, respectively. Both gaps are assumed to be complex at this stage. Unfortunately, the final expressions turns out to be very complicated as we shall see later. Our mean field Hamiltonian reads

$$H_M = \sum_{k>0} H_{Mk}$$

$$= \sum_{k>0} (\tilde{\xi}_k \sum_\sigma (n_{k\sigma} + n_{-k\sigma}) - \Delta_{Gk}(\alpha_k^* + \alpha_{-k}^*) - \Delta_{Gk}^*(\alpha_k + \alpha_{-k}) - 2\Delta_{gk}^*\beta_k - 2\Delta_{gk}\beta_k^* + C_k), \quad (3)$$

where

$$\alpha_k = a_{-k-}a_{k+} \quad , \quad \beta_k = b_{-k}b_k$$

$$C_k = \Delta_{Gk}\sigma_k^* + \Delta_{Gk}^*\sigma_k + \Delta_{gk}\tau_k^* + \Delta_{gk}^*\tau_k$$

$$\Delta_{Gp} := N^{-1}\sum_{k'} G_{kk'}\sigma_{k'} \quad , \quad \Delta_{gk} := N^{-1}\sum_{k'} g_{kk'}\tau_{k'} \quad (4)$$

with $\sigma_{Gk} = \dfrac{\text{Tr}\, e^{-\beta H_{Mk}}\alpha_k}{\text{Tr}\, e^{-\beta H_{Mk}}}$ and $\tau_{gk} = \dfrac{\text{Tr}\, e^{-\beta H_{Mk}}\beta_k}{\text{Tr}\, e^{-\beta H_{Mk}}}$ in practice. We followed the standard procedure of Bogolyubov et al. The details of the procedure can be found in [20] as well. The sum in the Hamiltonian H_M is over $\{k : k > 0\}$ denoting the set of all 1-fermion momenta restricted to a definite half-space of R^3.

Now it remains to diagonalize H_{Mk}. H_{Mk} acts in the 16-dimensional space M_k spanned by the vectors

$$|n_1 n_2 n_3 n_4\rangle := (a_{k+}^*)^{n_1}(a_{k-}^*)^{n_2}(a_{-k+}^*)^{n_3}(a_{-k-}^*)^{n_4}|0\rangle \quad (5)$$

where $n_i = 0, 1$ for $i = 1, 2, 3, 4$. The diagonalization can be done by the method presented in [21,22]. It consists in splitting M_k into invariant subspaces with fixed eigenvalues of the spin projection operator $2S_k$ and two seniorities $\Lambda_{k+}, \Lambda_{k-}$ defined as

$$2S_k = \sum_{\alpha=\pm 1}\sum_{\sigma=\pm 1} \sigma n_{\alpha k, \sigma}, \quad \Lambda_{k\sigma} = n_{k,\sigma} - n_{-k,-\sigma}, \quad \sigma = \pm \tag{6}$$

which commute with H_{Mk}, namely,

$$[H_{Mk}, 2S_k] = 0, \quad [H_{Mk}, \Lambda_{k,\sigma}] = 0. \tag{7}$$

M_k splits into nine such invariant subspaces M_{ki} ($i = 1, 2, ...9$) and due to the relations (7) diagonalization of H_{Mk} can be performed separately in each of them.

(A) There are four 1-dimensional common eigenspaces M_{ki} ($i = 1, 2, 3, 4$) of the operators (6) and H_{Mk}. They are spanned, respectively, by the following four vectors with the corresponding eigenvalues $2s, \lambda_+, \lambda_-, E_k$ of this operators equal as follows:

1. $|1010\rangle$ $2s = 2$ $\lambda_+ = 1$ $\lambda_- = -1$ $E_k = 2\tilde{\zeta}_k$
2. $|0101\rangle$ $2s = -2$ $\lambda_+ = -1$ $\lambda_- = 1$ $E_k = 2\tilde{\zeta}_k$
3. $|1100\rangle$ $2s = 0$ $\lambda_+ = 1$ $\lambda_- = 1$ $E_k = 2\tilde{\zeta}_k$
4. $|0011\rangle$ $2s = 0$ $\lambda_+ = -1$ $\lambda_- = -1$ $E_k = 2\tilde{\zeta}_k$

(B) There are also four 2-dimensional common eigenspaces M_{ki} ($i = 5, 6, 7, 8$) of $2S_k, \Lambda_{k\pm}$ and H_{Mk} spanned by the following pairs of vectors $|n_1 n_2 n_3 n_4\rangle$ and eigenvalues of these operators as follows:

5. $|1000\rangle$ $|1110\rangle$ $2s = 1$ $\lambda_+ = 1$ $\lambda_- = 0$ $E_{k\pm}$
6. $|0001\rangle$ $|0111\rangle$ $2s = -1$ $\lambda_+ = -1$ $\lambda_- = 0$ $E_{k\pm}$
7. $|0010\rangle$ $|1011\rangle$ $2s = 1$ $\lambda_+ = 0$ $\lambda_- = -1$ $E_{k\pm}$
8. $|0100\rangle$ $|1101\rangle$ $2s = -1$ $\lambda_+ = 0$ $\lambda_- = 1$ $E_{k\pm}$

where $E_{k\pm} = 2\tilde{\zeta}_k \pm E_{Gk}$ with $E_{Gk} = (\tilde{\zeta}_k^2 + |\Delta_{Gk}|^2)^{1/2}$. In each of the subspaces M_{ki} ($i = 5, 6, 7, 8$) the eigenproblem of H_{Mk} reduces to that of the matrix

$$\begin{pmatrix} \tilde{\zeta}_k & \Delta_{Gk}^* \\ \Delta_{Gk} & 3\tilde{\zeta}_k \end{pmatrix} \tag{8}$$

The eigenvectors of H_{Mk} in these subspaces have the form

$$|E_{k\pm}\rangle = c_{k\pm}|n_1 n_2 n_3 n_4\rangle + d_{k\pm}|m_1 m_2 m_3 m_4\rangle \tag{9}$$

where $\sum_{i=1}^{4} n_i = 1, \sum_{i=1}^{4} m_i = 3$ and

$$|c_{k\pm}|^2 = \frac{(\tilde{\zeta}_k \mp E_{Gk})^2}{|\Delta_{Gk}|^2 + (\tilde{\zeta}_k \mp E_{Gk})^2}, \quad |d_{k\pm}|^2 = \frac{|\Delta_{Gk}|^2}{|\Delta_{Gk}|^2 + (\tilde{\zeta}_k \mp E_{Gk})^2}$$

(C) There is one 4-dimensional common subspace M_{k9} of $2S_k, \Lambda_{k\pm}$ and H_{Mk} spanned by the vectors $|0000\rangle, |1001\rangle, |0110\rangle, |1111\rangle$. The eigenvalue $2S_k, \Lambda_{k\pm}$ in M_{k9} is $2s = \lambda_+ = \lambda_- = 0$. Let us use the following basis in M_{k9}:

$$\{|0000\rangle, -|1001\rangle, |0110\rangle, -|1111\rangle\} \tag{10}$$

and denote the projector on M_{k9} by P_{k9}. Therefore, in the basis (10) we have

$$P_{k9}H_k P_{k9} = \begin{pmatrix} 0 & \Delta_{Gk}^* & \Delta_{Gk}^* & 2\Delta_{gk}^* \\ \Delta_{Gk} & 2\tilde{\zeta}_k & 0 & \Delta_{Gk}^* \\ \Delta_{Gk} & 0 & 2\tilde{\zeta}_k & \Delta_{Gk}^* \\ 2\Delta_{gk} & \Delta_{Gk} & \Delta_{Gk} & 4\tilde{\zeta}_k \end{pmatrix} \quad (11)$$

Consecutively, the index **k** will be suppressed in this section. In terms of the unknown $x = E - 2\tilde{\zeta}$, the secular equation for the eigenvalues E of $P_{k9}H_k P_{k9}$ takes the form

$$(x)(x^3 + a_2 x^2 + a_1 x + a_0) = 0, \quad (12)$$

where

$a_0 = -4(\Delta_g \Delta_G^* \Delta_G^* + \Delta_g^* \Delta_G \Delta_G),$
$a_1 = -4(|\Delta_g|^2 + |\Delta_G|^2 + \tilde{\zeta}^2),$
$a_2 = 0.$

One root of Equation (12) is thus equal $x^{(2)} = 0$, yielding $E^{(2)} = 2\tilde{\zeta}$. The remaining three can be determined by passing to the standard form of a 3rd-order equation:

$$x^3 + 3px + 2q = 0 \quad (13)$$

with

$3p = a_1 = -4(|\Delta_g|^2 + |\Delta_G|^2 + \tilde{\zeta}^2)$
$2q = -4(\Delta_g \Delta_G^* \Delta_G^* + \Delta_g^* \Delta_G \Delta_G)$

Equation (13) has the solutions

$$y_k = 2(-1)^k \sqrt{r} \cos\left(\frac{\varphi}{3} + \frac{k\pi}{3}\right), \quad k = 0, \pm 1 \quad (14)$$

where $r = -p$, $\cos\varphi = tr^{-\frac{3}{2}}$ with $t = -q$. These solutions yield the corresponding eigenvalues of H_{Mk} in M_{k9}, viz.,

$$E^{(k)} = 2\tilde{\zeta} + y_k \quad (15)$$

Let $u^{(j)}, v_1^{(j)}, v_2^{(j)}, s^{(j)}$ denote the components of the eigenvectors $|E^{(j)}\rangle$ of $P_9 H P_9$ in the basis (10):

$$P_9 H P_9 |E^{(j)}\rangle = E^{(j)}|E^{(j)}\rangle \quad (16)$$

along with

$$|E^{(j)}\rangle = u^{(j)}|0000\rangle - v_1^{(j)}|1001\rangle + v_2^{(j)}|0110\rangle - s^{(j)}|1111\rangle, \quad j = 0, \pm 1, 2 \quad (17)$$

One finds

$$|E^{(2)}\rangle = |2\tilde{\zeta}\rangle = \frac{1}{\sqrt{2}}(|1001\rangle + |0110\rangle)$$

and the components of the remaining three eigenvectors $|E^{(j)}\rangle$ equal

$$|u^{(j)}|^2 = \frac{|a^{(j)}|^2}{D^{(j)}}, \quad v_1^{(j)} = v_2^{(j)}, \quad |v_1^{(j)}|^2 = \frac{|b^{(j)}|^2}{D^{(j)}}, \quad |s^{(j)}|^2 = \frac{|c^{(j)}|^2}{D^{(j)}}, \quad j = 0 \pm 1 \quad (18)$$

where

$$|a^{(i)}|^2 = (2\Delta_g^* \Delta_G - (4\tilde{\zeta} - E^{(i)})\Delta_G^*)(2\Delta_g \Delta_G^* - (4\tilde{\zeta} - E^{(i)})\Delta_G)(E^{(i)} - 2\tilde{\zeta})^2, \quad (19)$$

$$|b^{(i)}|^2 = 4(\Delta_g^* \Delta_G \Delta_G + \Delta_g \Delta_G^* \Delta_G^*) + |\Delta_G|^2 (E^{(i)} - 2\tilde{\zeta}))^2, \tag{20}$$

$$|c^{(i)}|^2 = (E^{(i)} \Delta_G + 2\Delta_g \Delta_G^*)(E^{(i)} \Delta_G^* + 2\Delta_g^* \Delta_G)(E^{(i)} - 2\tilde{\zeta})^2, \tag{21}$$

$$D^{(i)} = |a^{(i)}|^2 + 2|b^{(i)}|^2 + |c^{(i)}|^2 \tag{22}$$

4. The Case with the Zero BCS Potential

As it has been stated in Introduction, we are interested in some zero-temperature properties of H in which the BCS interaction is discarded. Especially, this concerns the electromagnetic response to a weak external magnetic field. The objective is to show how the electromagnetic kernel depends on the symmetry of a quartet order parameter and on the type of one-electron density of states in the band. Before we do that, we will give some details regarding the spectrum and ground state properties of such a system. The mean field Hamiltonian is

$$\overline{H}_g = \sum_{k>0} \left[\tilde{\zeta}_k \sum_\sigma (n_{k\sigma} + n_{-k\sigma}) - 2\Delta_{gk}(\beta_k + \beta_k^*) + C_k \right] = \sum_{k>0} \overline{H}_{gk}, \tag{23}$$

where Δ_{gk} is assumed to be real. Its eigenstructure has following form:

	eigenvector	H_k	2S	Λ_+	Λ_-
1.	$\|1000\rangle$	$\tilde{\zeta}_k$	1	1	0
2.	$\|0100\rangle$	$\tilde{\zeta}_k$	-1	0	1
3.	$\|0010\rangle$	$\tilde{\zeta}_k$	1	0	-1
4.	$\|0001\rangle$	$\tilde{\zeta}_k$	-1	-1	0
5.	$\|1010\rangle$	$2\tilde{\zeta}_k$	2	1	-1
6.	$\|0101\rangle$	$2\tilde{\zeta}_k$	-2	-1	1
7.	$\|1001\rangle$	$2\tilde{\zeta}_k$	0	0	0
8.	$\|0110\rangle$	$2\tilde{\zeta}_k$	0	0	0
9.	$\|1100\rangle$	$2\tilde{\zeta}_k$	0	1	1
10.	$\|0011\rangle$	$2\tilde{\zeta}_k$	0	-1	-1
11.	$\|1110\rangle$	$3\tilde{\zeta}_k$	1	1	0
12.	$\|0111\rangle$	$3\tilde{\zeta}_k$	-1	-1	0
13.	$\|1101\rangle$	$3\tilde{\zeta}_k$	-1	0	1
14.	$\|1011\rangle$	$3\tilde{\zeta}_k$	1	0	-1
15.	$u_k\|0000\rangle + v_k\|1111\rangle$	$2\tilde{\zeta}_k - 2E_k$	0	0	0
16.	$u_k\|1111\rangle - v_k\|0000\rangle$	$2\tilde{\zeta}_k + 2E_k$	0	0	0

It is worth noting that the vector number 15 is the ground state vector $|G_k\rangle$ of the system for momentum $\mathbf{k}$ and $E_k = \sqrt{\tilde{\zeta}_k^2 + \Delta_k^2}$, $u_k^2 = \frac{1}{2}(1 + \frac{\tilde{\zeta}_k}{E_k})$ and $v_k^2 = \frac{1}{2}(1 - \frac{\tilde{\zeta}_k}{E_k})$.

By making use of definition (4) we obtain the gap equation, namely

$$\Delta_{gk} = N^{-1} \frac{1}{2} \sum_{k'>0} g_{kk'} \langle G|\beta_{k'}|G\rangle = N^{-1} \sum_{k'} g_{kk'} \frac{\Delta_{gk'}}{2E_{k'}}. \tag{24}$$

The chemical potential μ can be found from the expression for the average number of particles per lattice site

$$n = N^{-1} \frac{1}{2} \sum_{k>0} \langle G|(n_{k+} + n_{k-} + n_{-k+} + n_{-k-})|G\rangle = \frac{2}{N} \sum_k v_k^2, \tag{25}$$

where $|G\rangle = \bigotimes_{k>0} |G_k\rangle$ is the total ground state vector. The ground state energy per lattice site reads

$$\frac{E_G}{N} = \frac{1}{N} \sum_k (\tilde{\xi}_k - E_k + \frac{\Delta_k^2}{2E_k}). \qquad (26)$$

The above equations will be used throughout this paper.

5. Interaction with External Electromagnetic Field at Zero Temperature

This section is to large extent based on perturbation theory used in the book by Abrikosov [23]. Our objective is to demonstrate the existence of the Meissner–Ochsenfeld effect in the system described by the Hamiltonian $H_f = H_g + H'$, where H' represents the perturbation due to a weak static external electromagnetic field described by the vector potential $\mathbf{A}(\mathbf{r})$. Thus

$$H' = \frac{1}{2m} \int d^3 r \psi^*(\mathbf{r}) \left[(-i\hbar \nabla - e\mathbf{A}/c)^2 - (-i\hbar \nabla)^2 \right] \psi(\mathbf{r}).$$

After making an expansion of the field operators $\psi^*(\mathbf{r}), \psi(\mathbf{r})$ one obtains

$$H' = H'_1 + H'_2$$

with

$$H'_1 = -\frac{e\hbar}{2mc} \sum_{\mathbf{k},\mathbf{q},\sigma} \mathbf{a}(\mathbf{q}) \cdot (\mathbf{q} + 2\mathbf{k}) a^*_{\mathbf{k}+\mathbf{q}\sigma} a_{\mathbf{k}\sigma}, \qquad (27)$$

and

$$H'_2 = \frac{e^2}{2mc^2} \sum_{\mathbf{k},\mathbf{q},\mathbf{q}',\sigma} \mathbf{a}(\mathbf{q}) \mathbf{a}(\mathbf{q}') a^*_{\mathbf{k}\sigma} a_{\mathbf{k}-\mathbf{q}-\mathbf{q}'\sigma}, \qquad (28)$$

where

$$\mathbf{a}(\mathbf{k}' - \mathbf{k}) = 1/N \int d^3 r \mathbf{A}(\mathbf{r}) \exp\left[-i(\mathbf{k}' - \mathbf{k}) \cdot \mathbf{r}\right].$$

Now we would like to find corrections to the ground state energy of the system E_G up to second order in $\mathbf{a}(\mathbf{q})$. E_G is connected with the eigenenergy $2\tilde{\xi}_k - 2E_k$, where $E_k = \sqrt{\tilde{\xi}_k^2 + \Delta_k^2}$. To this end, let us replace $\mathbf{q}$ with $\mathbf{k}' - \mathbf{k}$ in H'_1. In terms of new creation and annihilation operators a^*_{ki}, a_{ki} with $i = 1, 2, 3, 4$ (defined as $a^*_{k1} := a^*_{k+}, a^*_{k2} := a^*_{k-}, a^*_{k3} := a^*_{-k+}, a^*_{k4} := a^*_{-k-}$) H'_1 assumes the form

$$H'_1 = -\frac{e\hbar}{2mc} \sum_{\substack{k>0 \\ k'>0}} \Big\{ \mathbf{a}(\mathbf{k}' - \mathbf{k}, t) \cdot (\mathbf{k}' + \mathbf{k})(a^*_{k'1} a_{k1} - a^*_{k4} a_{k'4}) +$$
$$+ \mathbf{a}(-\mathbf{k}' - \mathbf{k}, t) \cdot (-\mathbf{k}' + \mathbf{k})(a^*_{k'3} a_{k1} - a^*_{k4} a_{k'2}) +$$
$$+ \mathbf{a}(\mathbf{k}' + \mathbf{k}, t) \cdot (\mathbf{k}' - \mathbf{k})(a^*_{k'1} a_{k3} - a^*_{k2} a_{k'4}) +$$
$$+ \mathbf{a}(-\mathbf{k}' + \mathbf{k}, t) \cdot (-\mathbf{k}' - \mathbf{k})(a^*_{k'3} a_{k3} - a^*_{k2} a_{k'2}) \Big\},$$

where $\{\mathbf{k} : \mathbf{k} > 0\}$ denotes the set of all 1-fermion momenta restricted to a definite half-space of $\mathbb{R}^3$. The expression above is a result of a rearrangement of terms entering (27) and switching $\mathbf{k}$ and $\mathbf{k}'$ in some terms. Thus, we are able to calculate the first order corrections, viz.,

$$E_1^{(1)} = \langle G|H'_1|G\rangle = \begin{cases} 0, & \mathbf{k}' = \mathbf{k}; \\ 0, & \mathbf{k}' \neq \mathbf{k}, \end{cases} \qquad (29)$$

and

$$E_2^{(1)} = \langle G|H'_2|G\rangle = \frac{e^2}{2mc^2} \sum_{\mathbf{k}\mathbf{q}\mathbf{q}'\sigma} \mathbf{a}(\mathbf{q}) \mathbf{a}(\mathbf{q}') \langle G|a^*_{\mathbf{k}\sigma} a_{\mathbf{k}-\mathbf{q}-\mathbf{q}'\sigma}|G\rangle = \frac{e^2}{2mc^2} \sum_{\mathbf{k}\mathbf{q}} 2v_k^2 \mathbf{a}(\mathbf{q}) \mathbf{a}(-\mathbf{q}). \qquad (30)$$

In (30) operators are in the original form, it is, with momentum indices belonging to whole momentum space. Now the calculation of the second order term $E_1^{(2)}$ can be performed, viz.,

$$E_1^{(2)} = \sum_{m \neq G} \frac{|\langle G|H_1'|m\rangle|^2}{E_G - E_m},$$

where $|m\rangle$ and E_m denote vectors representing excited states constructed from the vectors placed in Section 4 and their energies, respectively. E_G is the ground state energy. To get this correction the eigenstructure from Section 4 for momenta $\mathbf{k}$ and $\mathbf{k}'$ must be exploited. It is found that the only nonzero contributions are as follows:

$$|_{\mathbf{k}}\langle 1110|_{\mathbf{k}'}\langle 0001|a_{\mathbf{k}'1}^* a_{\mathbf{k}1}|G_{\mathbf{k}'}\rangle|G_{\mathbf{k}}\rangle|^2 = (u_{\mathbf{k}'} v_{\mathbf{k}})^2,$$

$$|_{\mathbf{k}}\langle 1000|_{\mathbf{k}'}\langle 0111|-a_{\mathbf{k}4}^* a_{\mathbf{k}'4}|G_{\mathbf{k}'}\rangle|G_{\mathbf{k}}\rangle|^2 = (u_{\mathbf{k}} v_{\mathbf{k}'})^2,$$

$$|_{\mathbf{k}}\langle 1110|_{\mathbf{k}'}\langle 0100|a_{\mathbf{k}'3}^* a_{\mathbf{k}1}|G_{\mathbf{k}'}\rangle|G_{\mathbf{k}}\rangle|^2 = (u_{\mathbf{k}'} v_{\mathbf{k}})^2,$$

$$|_{\mathbf{k}}\langle 1000|_{\mathbf{k}'}\langle 1101|-a_{\mathbf{k}4}^* a_{\mathbf{k}'2}|G_{\mathbf{k}'}\rangle|G_{\mathbf{k}}\rangle|^2 = (u_{\mathbf{k}} v_{\mathbf{k}'})^2,$$

$$|_{\mathbf{k}}\langle 1011|_{\mathbf{k}'}\langle 0001|a_{\mathbf{k}'1}^* a_{\mathbf{k}3}|G_{\mathbf{k}'}\rangle|G_{\mathbf{k}}\rangle|^2 = (u_{\mathbf{k}'} v_{\mathbf{k}})^2,$$

$$|_{\mathbf{k}}\langle 0010|_{\mathbf{k}'}\langle 0111|-a_{\mathbf{k}2}^* a_{\mathbf{k}'4}|G_{\mathbf{k}'}\rangle|G_{\mathbf{k}}\rangle|^2 = (u_{\mathbf{k}} v_{\mathbf{k}'})^2,$$

$$|_{\mathbf{k}}\langle 1011|_{\mathbf{k}'}\langle 0100|a_{\mathbf{k}'3}^* a_{\mathbf{k}3}|G_{\mathbf{k}'}\rangle|G_{\mathbf{k}}\rangle|^2 = (u_{\mathbf{k}'} v_{\mathbf{k}})^2,$$

$$|_{\mathbf{k}}\langle 0010|_{\mathbf{k}'}\langle 1101|-a_{\mathbf{k}2}^* a_{\mathbf{k}'2}|G_{\mathbf{k}'}\rangle|G_{\mathbf{k}}\rangle|^2 = (u_{\mathbf{k}} v_{\mathbf{k}'})^2,$$

where $u_{\mathbf{k}}^2 = \frac{1}{2}(1 + \frac{\xi_{\mathbf{k}}}{E_{\mathbf{k}}})$ and $v_{\mathbf{k}}^2 = \frac{1}{2}(1 - \frac{\xi_{\mathbf{k}}}{E_{\mathbf{k}}})$. After summing all nonzero contributions we obtain

$$E_1^{(2)} = \left(\frac{e\hbar}{2mc}\right)^2 \sum_{\mathbf{k}\mathbf{q}} \left(\frac{(u_{\mathbf{k}+\mathbf{q}} v_{\mathbf{k}})^2}{\xi_{\mathbf{k}+\mathbf{q}} - \xi_{\mathbf{k}} - 2(E_{\mathbf{k}} + E_{\mathbf{k}+\mathbf{q}})} + \frac{(v_{\mathbf{k}+\mathbf{q}} u_{\mathbf{k}})^2}{\xi_{\mathbf{k}} - \xi_{\mathbf{k}+\mathbf{q}} - 2(E_{\mathbf{k}} + E_{\mathbf{k}+\mathbf{q}})}\right) [\mathbf{a}(\mathbf{q}) \cdot (2\mathbf{k}+\mathbf{q})][\mathbf{a}(-\mathbf{q}) \cdot (2\mathbf{k}+\mathbf{q})]. \quad (31)$$

We exploited $\mathbf{a}^*(\mathbf{q}) = \mathbf{a}(-\mathbf{q})$ here. Furthermore the sum $\sum_{\substack{\mathbf{k}>0 \\ \mathbf{k}'>0}}$ was replaced with the unrestricted sum $\sum_{\mathbf{k}\mathbf{q}}$.

It is well known that the Hamiltonian of a system in an external magnetic field fulfils the following relation:

$$\delta H = -\frac{1}{c} \int \mathbf{j} \delta \mathbf{A} d^3 r,$$

where $\mathbf{j}$ is the current density operator. Averaging this relation over a given state, we obtain

$$\delta E = -\frac{1}{c} \int \langle \mathbf{j} \rangle \delta \mathbf{A} d^3 r. \quad (32)$$

Next we obtain expression (32) in Fourier representation, viz.,

$$\delta E = -\frac{N}{c} \sum_{\mathbf{q}} \mathbf{j}_{\mathbf{q}} \delta \mathbf{a}(-\mathbf{q}),$$

with

$$\mathbf{j}_{\mathbf{q}} = \frac{1}{N} \int d^3 r \mathbf{j} e^{i\mathbf{q}\cdot\mathbf{r}}.$$

In order to get the expression for j_q the corrections to the ground state energy are differentiated with respect to the vector potential. It yields

$$\delta E = 2\left[\frac{e^2}{mc^2}\sum_{kq} v_k^2 a(q) + \left(\frac{e\hbar}{2mc}\right)^2 \sum_{kq}\left(\frac{(u_{k+q}v_k)^2}{\tilde{\zeta}_{k+q} - \tilde{\zeta}_k - 2(E_k + E_{k+q})}\right.\right.$$
$$\left.\left.+ \frac{(v_{k+q}u_k)^2}{\tilde{\zeta}_k - \tilde{\zeta}_{k+q} - 2(E_k + E_{k+q})}\right)[a(q)\cdot(2k+q)](2k+q)\right]\delta a(-q). \tag{33}$$

Now having the expression above for δE at disposal the current density can be obtained,

$$j_q = -2\frac{e^2}{mcN}\sum_k v_k^2 a(q) - \frac{2}{cN}\left(\frac{e\hbar}{2m}\right)^2 \sum_k \left(\frac{(u_{k+q}v_k)^2}{\tilde{\zeta}_{k+q} - \tilde{\zeta}_k - 2(E_k + E_{k+q})}\right.$$
$$\left.+ \frac{(v_{k+q}u_k)^2}{\tilde{\zeta}_k - \tilde{\zeta}_{k+q} - 2(E_k + E_{k+q})}\right)[a(q)\cdot(2k+q)](2k+q). \tag{34}$$

For **q** tending to zero the relation for current density takes the following form

$$j_0 = -K(0,0)a(0), \tag{35}$$

with

$$K(0,0) = 2\frac{e^2}{mcN}\sum_k v_k^2 - \frac{8}{d}\frac{1}{cN}\left(\frac{e\hbar}{2m}\right)^2 \sum_k k^2 \frac{2(u_k v_k)^2}{4E_k},$$

where d is equal to the dimension of the system. The Meissner effect holds if $K(0,0) = \text{const} > 0$. This is known as the Schafroth's criterion. Two zeros that stand as the arguments in $K(0,0)$ correspond to static external field ($\omega = 0$) and $q = 0$. Finally, by making use of Equation (25) one obtains

$$K(0,0) = \frac{e^2}{mc}n - \frac{8}{d}\frac{1}{cN}\left(\frac{e\hbar}{2m}\right)^2 \sum_k k^2 \frac{2(u_k v_k)^2}{4E_k}. \tag{36}$$

The Equation (36) is the basis for the investigation of the response of the system under the study to a weak external magnetic field. We shall make it for three different symmetries of the order parameter: the pure s-wave pairing, the $d_{x^2-y^2}$-wave pairing and the mixed $s+id_{x^2+y^2}$-wave pairing cases. The p-wave case is not relevant to the problem of four-fermion potential in the form used in this paper because this type of pairing refers to the so-called equal-spin pairing but fermion quartets with the same $|k|$ cannot manifest the pairing of this kind of symmetry. In a more general case, when the pairs and the quartets are present in the system, one can consider combinations of different types of pairing including the p-wave pairing as well.

6. The Scenario with the Rectangular Density of States

At first, let us consider our system with the rectangular DOS, defined as:

$$\rho(\xi) = \begin{cases} \frac{1}{De}, & \xi \in (-\mu, De - \mu); \\ 0, & \text{otherwise,} \end{cases} \tag{37}$$

where De is the bandwidth. Let us begin with the order parameter with the pure s-wave pairing symmetry. In this case the coupling function $g_{kk'} = g > 0$ and as a consequence $\Delta_{gk} = \Delta_g$. In the thermodynamic limit the sums in Equations (24)–(26) and (36) turn into integrals, i.e.,

$$1 = \frac{g}{2De}\int_{-\mu}^{De-\mu}\frac{d\xi}{E} = \frac{g}{2De}\left(\text{arc sinh}\frac{De-\mu}{\Delta_g} + \text{arc sinh}\frac{\mu}{\Delta_g}\right), \tag{38}$$

$$n = \frac{1}{De}\int_{-\mu}^{De-\mu}d\xi\left(1-\frac{\xi}{E}\right) = 1 - \frac{1}{De}\left(\sqrt{(De-\mu)^2+\Delta_g^2}-\sqrt{\mu^2+\Delta_g^2}\right), \quad (39)$$

$$\begin{aligned}\frac{E_G}{N} &= \frac{1}{De}\int_{-\mu}^{De-\mu}d\xi\left(\xi-E+\frac{\Delta_g^2}{2E}\right) \\ &= \frac{1}{2}(De-2\mu) - \frac{1}{2De}\left((De-\mu)\sqrt{(De-\mu)^2+\Delta_g^2}+\mu\sqrt{\mu^2+\Delta_g^2}\right) - \frac{\Delta_g^2}{g}\end{aligned} \quad (40)$$

and finally

$$\begin{aligned}K(0,0) &= \frac{e^2}{mc}\left[n - \frac{1}{De}\int_{-\mu}^{De-\mu}d\xi(\xi+\mu)\frac{\Delta_g^2}{4E^3}\right] \\ &= \frac{e^2}{mc}\left[n - \frac{\Delta_g^2}{4De}\left[\frac{1}{\sqrt{\mu^2+\Delta_g^2}}-\frac{1}{\sqrt{(De-\mu)^2+\Delta_g^2}}+\frac{\mu}{\Delta_g^2}\left(\frac{De-\mu}{\sqrt{(De-\mu)^2+\Delta_g^2}}+\frac{\mu}{\sqrt{\mu^2+\Delta_g^2}}\right)\right]\right],\end{aligned} \quad (41)$$

where $E = \sqrt{\xi^2+\Delta_g^2}$. The one-electron dispersion relation counted with respect to μ is as follows $\xi_k = \frac{\hbar^2 k^2}{2m} - \mu$ while the system is two-dimensional. The Equations (38)–(40) are the same as in the BCS model with the rectangular DOS; However, the Equation (41) differs from the BCS counterpart that has the following form:

$$K(0,0) = \frac{e^2 n}{mc}.$$

As one can notice, at the zero temperature, there is only the so-called diamagnetic part in the electromagnetic kernel. At higher temperatures, the paramagnetic term appears and increases up to the transition temperature T_c. At this temperature, both terms cancel each other, which means the transition of a superconducting system goes back to the normal state. This happens so independently of the symmetry of pairing. The situation is significantly different if we replace the BCS interaction with the four-fermion attraction. By the appearance of the paramagnetic term, the Meissner effect gets weaker even at zero temperature. This was discovered in 2004 [24]. In that paper, the electromagnetic kernel at zero temperature reads

$$K(0,0) = \frac{3}{4}\frac{e^2 n}{mc}. \quad (42)$$

Now, for simplicity sake, let us consider the half-filled band case, i.e., $n=1$. Then the Equations (38)–(41) yield $\mu = \frac{De}{2}$ and

$$\frac{E_G}{N} = -\frac{1}{2}\sqrt{\frac{De^2}{4}+\Delta_g^2} - \frac{\Delta_g^2}{g}, \quad \Delta_g = \frac{\frac{De}{2}}{\sinh\frac{De}{g}}, \quad K(0,0) = \frac{e^2}{mc}\left[1-\frac{1}{8}\frac{De}{\sqrt{\frac{De^2}{4}+\Delta_g^2}}\right], \quad (43)$$

Note that in the limit of the shrinking band $De \to 0$ (the atomic limit) one obtains the BCS result $K(0,0) = \frac{e^2}{mc}$. In the opposite regime $\Delta_g \ll \frac{De}{2}$ we obtain the result (42). This means that the external magnetic field penetrates a superconductor deeper in the case of the system with the four-fermion attraction and the wider band than in the case of the system with the BCS potential or than with the four-fermion potential but with a very narrow band.

Now, let us deal with the scenario where $d_{x^2-y^2}$-wave pairing symmetry. Let us take the coupling function and the order parameter in the form $g_{kk'} = g(k_x^2-k_y^2)(k_x'^2-k_y'^2)$ and $\Delta_{gk} = \Delta_g(k_x^2-k_y^2)$, respectively. In angle representation one gets $g(\phi,\phi') = g\cos 2\phi \cos 2\phi'$ and $\Delta_g(\phi) = \Delta_g \cos 2\phi$. The Equations (24)–(26) and (36) take the following form:

$$1 = \frac{g}{4\pi De}\int_0^{2\pi}d\phi \cos^2 2\phi\left(\operatorname{arc sinh}\frac{De-\mu}{\Delta_g|\cos 2\phi|}+\operatorname{arc sinh}\frac{\mu}{\Delta_g|\cos 2\phi|}\right), \quad (44)$$

$$n = 1 - \frac{1}{2\pi De} \int_0^{2\pi} d\phi \left(\sqrt{(De - \mu)^2 + \Delta_g^2 \cos^2 2\phi} - \sqrt{\mu^2 + \Delta_g^2 \cos^2 2\phi} \right), \tag{45}$$

$$\frac{E_G}{N} = \frac{1}{2}(De - 2\mu) - \frac{1}{4\pi De} \int_0^{2\pi} d\phi \left((De - \mu)\sqrt{(De - \mu)^2 + \Delta_g^2 \cos^2 2\phi} + \mu\sqrt{\mu^2 + \Delta_g^2 \cos^2 2\phi} \right) - \frac{\Delta_g^2}{g} \tag{46}$$

and finally

$$K(0,0) = \frac{e^2}{mc} \left[n - \frac{\Delta_g^2}{8\pi De} \left[\int_0^{2\pi} d\phi \cos^2 2\phi \left(\frac{1}{\sqrt{\mu^2 + \Delta_g^2 \cos^2 2\phi}} - \frac{1}{\sqrt{(De - \mu)^2 + \Delta_g^2 \cos^2 2\phi}} \right) + \right. \right.$$
$$\left. \left. + \frac{\mu}{\Delta_g^2} \int_0^{2\pi} d\phi \left(\frac{De - \mu}{\sqrt{(De - \mu)^2 + \Delta_g^2 \cos^2 2\phi}} + \frac{\mu}{\sqrt{\mu^2 + \Delta_g^2 \cos^2 2\phi}} \right) \right] \right]. \tag{47}$$

In the case of the half-filled band ($n = 1$) the Equations (44)–(47) take a simpler form, namely:

$$\mu = \frac{De}{2}, \quad 1 = \frac{g}{2\pi De} \int_0^{2\pi} d\phi \cos^2 2\phi \, \text{arc sinh} \frac{De}{2\Delta_g |\cos 2\phi|}, \tag{48}$$

$$\frac{E_G}{N} = -\frac{1}{4\pi} \int_0^{2\pi} d\phi \sqrt{\frac{De^2}{4} + \Delta_g^2 \cos^2 2\phi} - \frac{\Delta_g^2}{g} \tag{49}$$

and

$$K(0,0) = \frac{e^2}{mc} \left[1 - \frac{De}{16\pi} \int_0^{2\pi} \frac{d\phi}{\sqrt{\frac{De^2}{4} + \Delta_g^2 \cos^2 2\phi}} \right]. \tag{50}$$

The integrals in the above equations can be expressed by the elliptic integrals which can be calculated numerically what is made later in the work. One can try to obtain some analytical results, namely, assume that $\Delta_g \ll \frac{De}{2}$ then one can apply the following expansions:

$$\sqrt{\frac{De^2}{4} + \Delta_g^2 \cos^2 2\phi} = \frac{De}{2} + \frac{\Delta_g^2 \cos^2 2\phi}{De} + ...,$$

$$\frac{1}{\sqrt{\frac{De^2}{4} + \Delta_g^2 \cos^2 2\phi}} = \frac{2}{De} - \frac{4}{De^3} \Delta_g^2 \cos^2 2\phi + ...$$

If we limit ourselves to the first order terms and substitute them into the Equations (49) and (50) we will obtain after integration:

$$\frac{E_G}{N} \approx -\frac{1}{4} De - \frac{1}{4} \frac{\Delta_g^2}{De} - \frac{\Delta_g^2}{g} \tag{51}$$

and

$$K(0,0) \approx \frac{e^2}{mc} \left[\frac{3}{4} + \frac{1}{4} \frac{\Delta_g^2}{De^2} \right]. \tag{52}$$

Notice that the neglecting of the second term in (52) gives the same result as (42). At last let us deal with the mixed $s + id_{x^2-y^2}$-wave pairing symmetry. We take the coupling function in the form $g_{kk'} = g_0 + g_1(k_x^2 - k_y^2)(k_x'^2 - k_y'^2)$, where g_0 and g_1 are positive real numbers. The gap parameter is complex and reads $\Delta_{gk} = \Delta_s + i\Delta_d(k_x^2 - k_y^2)$ and the equation for it at zero temperature is as follows

$$\Delta_{gk} = \frac{1}{N} \sum_{k'} \frac{\Delta_{gk'}}{2E_{k'}},$$

where $E_\mathbf{k} = \sqrt{\zeta_\mathbf{k} + \Delta_s^2 + \Delta_d^2(k_x^2 - k_y^2)^2}$. This equation splits up to two coupled equations for Δ_s and Δ_d, namely,

$$1 = \frac{g_0}{N}\sum_\mathbf{k}\frac{1}{2E_\mathbf{k}}, \quad 1 = \frac{g_1}{N}\sum_\mathbf{k}\frac{(k_x^2 - k_y^2)^2}{2E_\mathbf{k}}.$$

In the angle representation the factor $k_x^2 - k_y^2$ is replaced by $\cos 2\phi$. In this representation our problem can be expressed by the set of equations:

$$1 = \frac{g_0}{4\pi De}\int_0^{2\pi} d\phi \left(\text{arc sinh}\frac{De - \mu}{\sqrt{\Delta_s^2 + \Delta_d^2 \cos^2 2\phi}} + \text{arc sinh}\frac{\mu}{\sqrt{\Delta_s^2 + \Delta_d^2 \cos^2 2\phi}}\right), \quad (53)$$

$$1 = \frac{g_1}{4\pi De}\int_0^{2\pi} d\phi \cos^2 2\phi \left(\text{arc sinh}\frac{De - \mu}{\sqrt{\Delta_s^2 + \Delta_d^2 \cos^2 2\phi}} + \text{arc sinh}\frac{\mu}{\sqrt{\Delta_s^2 + \Delta_d^2 \cos^2 2\phi}}\right), \quad (54)$$

$$n = 1 - \frac{1}{2\pi De}\int_0^{2\pi} d\phi \left(\sqrt{(De - \mu)^2 + \Delta_s^2 + \Delta_d^2 \cos^2 2\phi} - \sqrt{\mu^2 + \Delta_s^2 + \Delta_d^2 \cos^2 2\phi}\right), \quad (55)$$

$$\frac{E_G}{N} = \frac{1}{2}(De - 2\mu) - \frac{\Delta_d^2}{g_1} - \frac{\Delta_s^2}{g_0} -$$

$$- \frac{1}{4\pi De}\int_0^{2\pi} d\phi \left((De - \mu)\sqrt{(De - \mu)^2 + \Delta_s^2 + \Delta_d^2 \cos^2 2\phi} + \mu\sqrt{\mu^2 + \Delta_s^2 + \Delta_d^2 \cos^2 2\phi}\right) \quad (56)$$

and

$$K(0,0) = \frac{e^2}{mc}\left[n - \frac{1}{8\pi De}\left[\int_0^{2\pi} d\phi(\Delta_s^2 + \Delta_d^2 \cos^2 2\phi)\left(\frac{1}{\sqrt{\mu^2 + \Delta_s^2 + \Delta_d^2 \cos^2 2\phi}} - \frac{1}{\sqrt{(De-\mu)^2 + \Delta_s^2 + \Delta_d^2 \cos^2 2\phi}}\right) + \right.\right.$$
$$\left.\left. + \mu\int_0^{2\pi} d\phi \left(\frac{De - \mu}{\sqrt{(De-\mu)^2 + \Delta_s^2 + \Delta_d^2 \cos^2 2\phi}} + \frac{\mu}{\sqrt{\mu^2 + \Delta_s^2 + \Delta_d^2 \cos^2 2\phi}}\right)\right]\right]. \quad (57)$$

Because we are interested in the half-filled band case the above equations take a much simpler form, i.e., $\mu = \frac{De}{2}$ and

$$1 = \frac{g_0}{2\pi De}\int_0^{2\pi} \text{arc sinh}\frac{De}{2\sqrt{\Delta_s^2 + \Delta_d^2 \cos^2 2\phi}} d\phi, \quad (58)$$

$$1 = \frac{g_1}{2\pi De}\int_0^{2\pi} \cos^2 2\phi \, \text{arc sinh}\frac{De}{2\sqrt{\Delta_s^2 + \Delta_d^2 \cos^2 2\phi}} d\phi, \quad (59)$$

$$\frac{E_G}{L} = -\frac{1}{4\pi}\int_0^{2\pi} d\phi \sqrt{\frac{De^2}{4} + \Delta_s^2 + \Delta_d^2 \cos^2 2\phi} - \frac{\Delta_s^2}{g_0} - \frac{\Delta_d^2}{g_1} \quad (60)$$

and finally

$$K(0,0) = \frac{e^2}{mc}\left[1 - \frac{De}{16\pi}\int_0^{2\pi} d\phi \frac{1}{\sqrt{\frac{De^2}{4} + \Delta_s^2 + \Delta_d^2 \cos^2 2\phi}}\right]. \quad (61)$$

Once again, let us try to find some analytical expressions for $\frac{E_G}{N}$ and $K(0,0)$. Note that in the limit of $De \to 0$ one obtains the BCS result $K(0,0) = \frac{e^2}{mc}$. Therefore, the narrower the band is the more similar to the BCS superconductor the system is. Reversely, for the sufficiently wide band, i.e., $\frac{De}{2}, \Delta_s \gg \Delta_d$ one has the following expansions:

$$\sqrt{\frac{De^2}{4} + \Delta_s^2 + \Delta_d^2 \cos^2 2\phi} = \sqrt{\frac{De^2}{4} + \Delta_s^2} + \frac{1}{2}\frac{\Delta_d^2 \cos^2 2\phi}{\sqrt{\frac{De^2}{4} + \Delta_s^2}} + \dots,$$

$$\frac{1}{\sqrt{\frac{De^2}{4} + \Delta_s^2 + \Delta_d^2 \cos^2 2\phi}} = \frac{1}{\sqrt{\frac{De^2}{4} + \Delta_s^2}} - \frac{1}{2}\frac{1}{(\frac{De^2}{4} + \Delta_s^2)^{\frac{3}{2}}}\Delta_d^2 \cos^2 2\phi + \dots$$

It suffices to limit ourselves to the zero and first order terms and substitute them into Equations (60) and (61). After integrating we obtain

$$\frac{E_G}{N} \approx -\frac{1}{2}\sqrt{\frac{De^2}{4} + \Delta_s^2} - \frac{1}{8}\frac{\Delta_d^2}{\sqrt{\frac{De^2}{4} + \Delta_s^2}} - \frac{\Delta_d^2}{g_1} - \frac{\Delta_s^2}{g_0} \tag{62}$$

and finally

$$K(0,0) \approx \frac{e^2}{mc}\left[1 - \frac{1}{8}\frac{De}{\sqrt{\frac{De^2}{4} + \Delta_s^2}}\left(1 - \frac{1}{4}\frac{\Delta_d^2}{\frac{De^2}{4} + \Delta_s^2}\right)\right]. \tag{63}$$

The last equation can be simplified if we neglect the term proportional to Δ_d^2 because it is much lesser than the unity. Therefore, this yields

$$K(0,0) \approx \frac{e^2}{mc}\left[1 - \frac{1}{8}\frac{De}{\sqrt{\frac{De^2}{4} + \Delta_s^2}}\right]. \tag{64}$$

The kernel (64) fulfills the double inequality

$$\frac{3}{4}\frac{e^2}{mc} < K(0,0) < \frac{e^2}{mc}. \tag{65}$$

We have analyzed the situation in which the component with the s-wave pairing symmetry and the bandwidth are much greater than the $d_{x^2-y^2}$-wave component. If we admit that $\frac{De}{2} \ll \Delta_s, \Delta_d$ then the ground state energy and the electromagnetic kernel read

$$\frac{E_G}{N} \approx -\frac{1}{4}De - \frac{1}{2}\frac{\Delta_s^2}{De} - \frac{1}{4}\frac{\Delta_d^2}{De} - \frac{\Delta_d^2}{g_1} - \frac{\Delta_s^2}{g_0} \tag{66}$$

and

$$K(0,0) \approx \frac{e^2}{mc}\left[\frac{3}{4} + \frac{1}{2}\frac{\Delta_s^2}{De^2} + \frac{1}{4}\frac{\Delta_d^2}{De^2}\right]. \tag{67}$$

Looking at the Equation (67) one can notice that the inequality (65) holds.

7. The Scenario with the Tight-Binding Model—The One-Dimensional Case

In this section we shall do the same kind of calculations but instead of the less realistic model with rectangular density of states we shall consider the one-dimensional model in the frame of the tight-binding approximation. In the one-dimensional system the dispersion relation has the form $\zeta_k = -2t \cos k - \mu$. As is known in one-dimensional systems due to strong quantum fluctuations there is no long-range order even at $T = 0$ but one can assume that the investigated chain interacts with environment that stabilizes the superconductivity in the chain. One can say that we have to deal with the effective model. In the one-dimensional case one can take under considerations the s-wave pairing only. It is no use to investigate the $d_{x^2-y^2}$-wave pairing case because the momentum has only one component. The gap equation reads

$$1 = \frac{g}{2\pi} \int_0^{\pi} \frac{dk}{\sqrt{(2t\cos k + \mu)^2 + \Delta_g^2}} = \frac{g}{2\pi} \int_{-2t-\mu}^{2t-\mu} \frac{d\xi}{\sqrt{\xi^2 + \Delta_g^2} \sqrt{4t^2 - (\xi+\mu)^2}}, \quad (68)$$

where after the introducing of the density of states this equation has been easily transformed to the second form. Note that at half filling ($\mu = 0$) DOS has two the so-called Van Hove singularities at $\xi = \pm 2t$. Next, the equation for the average number of particles per lattice site is

$$n = 1 + \frac{1}{\pi} \int_0^{\pi} dk \frac{2t\cos k + \mu}{\sqrt{(2t\cos k + \mu)^2 + \Delta_g^2}} = 1 - \frac{1}{\pi} \int_{-2t-\mu}^{2t-\mu} \frac{d\xi}{\sqrt{4t^2 - (\xi+\mu)^2}} \frac{\xi}{\sqrt{\xi^2 + \Delta_g^2}}. \quad (69)$$

The ground state energy per lattice site is expressed via

$$\frac{E_G}{N} = -\mu - \frac{1}{\pi} \int_0^{\pi} dk \sqrt{(2t\cos k + \mu)^2 + \Delta_g^2} + \frac{\Delta_g^2}{g} = -\mu - \frac{1}{\pi} \int_{-2t-\mu}^{2t-\mu} \frac{\sqrt{\xi^2 + \Delta_g^2}}{\sqrt{4t^2 - (\xi+\mu)^2}} d\xi + \frac{\Delta_g^2}{g}. \quad (70)$$

Finally, the electromagnetic kernel can be found from

$$K(0,0) = \frac{e^2}{mc} \left[n - \frac{t\Delta_g^2}{2\pi} \int_0^{\pi} \frac{k^2 dk}{((2t\cos k + \mu)^2 + \Delta_g^2)^{3/2}} \right]$$

$$= \frac{e^2}{mc} \left[n - \frac{t\Delta_g^2}{2\pi} \int_{-2t-\mu}^{2t-\mu} d\xi \frac{\left(\arccos\left(-\frac{\xi+\mu}{2t}\right)\right)^2}{\sqrt{4t^2 - (\xi+\mu)^2}} \frac{1}{(\xi^2 + \Delta_g^2)^{3/2}} \right], \quad (71)$$

where $t = \frac{\hbar^2}{2ma^2}$ is the hopping parameter while a is the lattice constant that is put here equal the unity. In this section we are interested in the half-filled case ($n = 1$) as well. The above equations get a simpler form. Let us start from the chemical potential $\mu = 0$ and the gap equation, namely,

$$1 = \frac{g}{\pi} \int_0^{2t} d\xi \frac{1}{\sqrt{4t^2 - \xi^2}} \frac{1}{\sqrt{\xi^2 + \Delta_g^2}}. \quad (72)$$

Note that in the one-dimensional case the bandwidth $De = 4t$. The ground state energy per lattice site is as follows

$$\frac{E_G}{N} = -\frac{2}{\pi} \int_0^{2t} d\xi \frac{\sqrt{\xi^2 + \Delta_g^2}}{\sqrt{4t^2 - \xi^2}} + \frac{\Delta_g^2}{g}. \quad (73)$$

Finally, the electromagnetic kernel reads

$$K(0,0) = \frac{e^2}{mc} \left[1 - \frac{t\Delta_g^2}{2\pi} \int_{-2t}^{2t} d\xi \frac{\left(\arccos\left(-\frac{\xi}{2t}\right)\right)^2}{\sqrt{4t^2 - \xi^2}} \frac{1}{(\xi^2 + \Delta_g^2)^{3/2}} \right]. \quad (74)$$

We can try to find an analytical solution of the gap Equation (72) in the $\frac{De}{2} \ll \Delta_g$ regime (the strong coupling regime). In this case the expansion of $\frac{1}{\sqrt{4t^2 \cos^2 k + \Delta_g^2}} \approx \frac{1}{\Delta_g} - \frac{1}{2} \frac{4t^2 \cos^2 k}{\Delta_g^3}$ to the first order is useful, then the Equation (72) reduces to a cubic equation that can be solved by the method from Section 3. This equation takes the form

$$\Delta_g^3 - \frac{g}{2}\Delta_g^2 + \frac{gt^2}{2} = 0. \qquad (75)$$

After finding the proper solution of the Equation (75) one can make use of the similar expansions of $\sqrt{4t^2\cos^2 k + \Delta_g^2} \approx \Delta_g + \frac{1}{2}\frac{4t^2\cos^2 k}{\Delta_g}$ and $\frac{1}{(4t^2\cos^2 k+\Delta_g^2)^{3/2}} \approx \frac{1}{\Delta_g^3} - \frac{3}{2}\frac{4t^2\cos^2 k}{\Delta_g^5}$ and use them for the calculation of the ground state energy and the electromagnetic kernel, i.e.,

$$\frac{E_G}{N} \approx -\Delta_g - \frac{t^2}{\Delta_g} + \frac{\Delta_g^2}{g} \qquad (76)$$

and

$$K(0,0) \approx \frac{e^2}{mc}\left[1 - \frac{\pi^2}{6}\frac{t}{\Delta_g} + \left(2\pi + \frac{3}{4}\right)\frac{t^3}{\Delta_g^3}\right]. \qquad (77)$$

The last equation can be approximated by the neglecting of the third term in the square brackets, i.e.,

$$K(0,0) \approx \frac{e^2}{mc}\left[1 - \frac{\pi^2}{6}\frac{t}{\Delta_g}\right]. \qquad (78)$$

It is visible, that the paramagnetic term is proportional to the hopping parameter t. The lesser it is the greater the kernel becomes. In the limit $t \to 0$ the kernel tends to the BCS one. As one can convince oneself that analytical result (78) is in agreement with numerical calculation outcomes which can be found in Section 9.

8. The Scenario with the Tight-Binding Model—The Two-Dimensional Case

In this section we shall do the same kind of calculations but we shall consider the two-dimensional model in the frame of the tight-binding approximation. In the case of two-dimensional one we shall pay an attention on the simple square lattice with the dispersion relation counted with respect to μ, i.e., $\check{\zeta}_k = -2t(\cos k_x + \cos k_y) - \mu$, where t is the hopping parameter while the lattice constant is put equal to the unity. Again, we are interested in the study of the following symmetries of the gaps: the s-wave pairing and the mixed $s + id_{x^2-y^2}$-wave pairing ones. The pure $d_{x^2-y^2}$-wave pairing symmetry case is treated in a numerical manner. The expressions for this case can be obtained from the equations for the $s + id_{x^2-y^2}$-wave pairing symmetry case by putting $\Delta_s = 0$ in them.

Let us start with the gap equation for Δ_g with the s-wave pairing symmetry. This reads

$$1 = \frac{g}{2\pi^2}\int_0^\pi dk_x \int_0^\pi dk_y \frac{1}{\sqrt{(2t(\cos k_x + \cos k_y) + \mu)^2 + \Delta_g^2}}, \qquad (79)$$

next the equation for the average number of particles per lattice site

$$n = 1 + \frac{1}{\pi^2}\int_0^\pi dk_x \int_0^\pi dk_y \frac{2t(\cos k_x + \cos k_y) + \mu}{\sqrt{(2t(\cos k_x + \cos k_y) + \mu)^2 + \Delta_g^2}} \qquad (80)$$

and those for the ground state energy per lattice site and the electromagnetic kernel:

$$\frac{E_G}{N} = -\mu - \frac{1}{\pi^2}\int_0^\pi dk_x \int_0^\pi dk_y \sqrt{(2t(\cos k_x + \cos k_y) + \mu)^2 + \Delta_g^2} + \frac{\Delta_g^2}{g}, \qquad (81)$$

$$K(0,0) = \frac{e^2}{mc}\left[n - \frac{t\Delta_g^2}{4\pi^2}\int_0^\pi dk_x \int_0^\pi dk_y (k_x^2 + k_y^2)\frac{1}{((2t(\cos k_x + \cos k_y) + \mu)^2 + \Delta_g^2)^{\frac{3}{2}}}\right]. \quad (82)$$

These equations for the half-filled band case can be produced by putting $n = 1$ and $\mu = 0$ and substituting these values into them. Next, in order to find some analytical results we follow the procedure used in the previous section, i.e., assume that $\frac{De}{2} \ll \Delta_g$, where $De = 8t$ is the bandwidth in the two-dimensional system. We will exploit the following expansions to the first order terms:

$$\sqrt{4t^2(\cos k_x + \cos k_y)^2 + \Delta_g^2} \approx \Delta_g + \frac{1}{2}\frac{4t^2(\cos k_x + \cos k_y)^2}{\Delta_g},$$

$$\frac{1}{\sqrt{4t^2(\cos k_x + \cos k_y)^2 + \Delta_g^2}} \approx \frac{1}{\Delta_g} - \frac{1}{2}\frac{4t^2(\cos k_x + \cos k_y)^2}{\Delta_g^3}$$

and

$$\frac{1}{(4t^2(\cos k_x + \cos k_y)^2 + \Delta_g^2)^{\frac{3}{2}}} \approx \frac{1}{\Delta_g^3} - \frac{3}{2}\frac{4t^2(\cos k_x + \cos k_y)^2}{\Delta_g^5}.$$

These expansions are now substituted into the Equations (79), (81) and (82) and after making the integration one obtains:

$$1 = \frac{g}{2}\left[\frac{1}{\Delta_g} - 2\frac{t^2}{\Delta_g^3}\right] \Leftrightarrow \Delta_g^3 - \frac{g}{2}\Delta_g^2 + gt^2 = 0, \quad (83)$$

$$\frac{E_G}{N} \approx -\Delta_g - 2\frac{t^2}{\Delta_g} + \frac{\Delta_g^2}{g} \quad (84)$$

and

$$K(0,0) \approx \frac{e^2}{mc}\left[1 - \frac{\pi^2}{6}\frac{t}{\Delta_g} + \frac{3}{2}\frac{t^3}{\Delta_g^3}\left(\frac{1}{2} + \frac{4}{3}\pi + \frac{\pi^2}{3}\right)\right].$$

If $\frac{t}{\Delta_g} \ll 1$ then we can approximate $K(0,0)$ by

$$K(0,0) \approx \frac{e^2}{mc}\left[1 - \frac{\pi^2}{6}\frac{t}{\Delta_g}\right]. \quad (85)$$

Note that (85) is the same as (78).

The pure $d_{x^2-y^2}$-wave pairing case will be investigated in another paper. We will look now at the case with mixed symmetry of pairing, namely, the $s + id_{x^2-y^2}$. The order parameter has the form $\Delta_{gk} = \Delta_s + i\Delta_d(\cos k_x - \cos k_y)$ while its module $|\Delta_{gk}| = \sqrt{\Delta_s^2 + \Delta_d^2(\cos k_x - \cos k_y)^2}$. The coupling function is $g_{kk'} = g_0 + g_1(\cos k_x - \cos k_y)(\cos k'_x - \cos k'_y)$, where g_0, g_1 are real positive numbers. Similarly, as it was in Section 6, we shall give the fundamental equations for our problem: the gap equations, the equation for the chemical potential, the ground state energy and the electromagnetic kernel. They read

$$1 = \frac{g_0}{2\pi^2}\int_0^\pi dk_x \int_0^\pi dk_y \frac{1}{\sqrt{(2t(\cos k_x + \cos k_y) + \mu)^2 + \Delta_s^2 + \Delta_d^2(\cos k_x - \cos k_y)^2}}, \quad (86)$$

$$1 = \frac{g_1}{2\pi^2}\int_0^\pi dk_x \int_0^\pi dk_y \frac{(\cos k_x - \cos k_y)^2}{\sqrt{(2t(\cos k_x + \cos k_y) + \mu)^2 + \Delta_s^2 + \Delta_d^2(\cos k_x - \cos k_y)^2}}, \quad (87)$$

$$n = 1 + \frac{1}{\pi^2} \int_0^\pi dk_x \int_0^\pi dk_y \frac{2t(\cos k_x + \cos k_y) + \mu}{\sqrt{(2t(\cos k_x + \cos k_y) + \mu)^2 + \Delta_s^2 + \Delta_d^2(\cos k_x - \cos k_y)^2}}, \qquad (88)$$

$$\frac{E_G}{N} = -\mu - \frac{1}{\pi^2} \int_0^\pi dk_x \int_0^\pi dk_y \sqrt{(2t(\cos k_x + \cos k_y) + \mu)^2 + \Delta_s^2 + \Delta_d^2(\cos k_x - \cos k_y)^2} + \frac{\Delta_s^2}{g_0} + \frac{\Delta_d^2}{g_1} \qquad (89)$$

and

$$K(0,0) = \frac{e^2}{mc}\left[n - \frac{t}{4\pi^2}\int_0^\pi dk_x \int_0^\pi dk_y (k_x^2 + k_y^2) \frac{\Delta_s^2 + \Delta_d^2(\cos k_x - \cos k_y)^2}{((2t(\cos k_x + \cos k_y) + \mu)^2 + \Delta_s^2 + \Delta_d^2(\cos k_x - \cos k_y)^2)^{\frac{3}{2}}}\right]. \qquad (90)$$

Because we are interested in the investigation of the half-filled band case it suffices to put $n = 1$. For this number of electrons one obtains the chemical potential $\mu = 0$. In order to obtain those equations for the half-filled band case, one needs to substitute these two values into them. As has been already done for the pure s-wave pairing case we can make some analytical calculations. Let us assume that $\frac{D_e}{2} \ll \Delta_s$ and $\Delta_d \ll \Delta_s$. Again let us make use of the following expansions to the first order terms:

$$\sqrt{4t^2(\cos k_x + \cos k_y)^2 + \Delta_s^2 + \Delta_d^2(\cos k_x - \cos k_y)^2} \approx$$

$$\Delta_s + \frac{1}{2}\frac{4t^2(\cos k_x + \cos k_y)^2 + \Delta_d^2(\cos k_x - \cos k_y)^2}{\Delta_s}$$

and

$$\frac{1}{(4t^2(\cos k_x + \cos k_y)^2 + \Delta_s^2 + \Delta_d^2(\cos k_x - \cos k_y)^2)^{\frac{3}{2}}} \approx$$

$$\frac{1}{\Delta_s^3} - \frac{3}{2}\frac{4t^2(\cos k_x + \cos k_y)^2 + \Delta_d^2(\cos k_x - \cos k_y)^2}{\Delta_s^5}.$$

Those expansions are next substituted into the integrals in (89) and (90). After the integration, one obtains the expressions for the ground state energy per lattice site together with the expression for the electromagnetic kernel, namely,

$$\frac{E_G}{N} \approx -\Delta_s - \frac{4t^2 + \Delta_d^2}{2\Delta_s} + \frac{\Delta_s^2}{g_0} + \frac{\Delta_d^2}{g_1} \qquad (91)$$

and

$$K(0,0) = \frac{e^2}{mc}\left[1 - \frac{1}{4\pi^2}\frac{t}{\Delta_s}\left[\frac{2}{3}\pi^4 - \frac{3}{2}\frac{4t^2 + \Delta_d^2}{\Delta_s^2}\left(\frac{4}{3}\pi^3 + \frac{1}{3}\pi^4 + \frac{\pi^2}{2}\right)\right] - \right. \qquad (92)$$
$$\left. - \frac{1}{4\pi^2}\frac{t}{\Delta_s}\frac{\Delta_d^2}{\Delta_s^2}\left[\frac{4}{3}\pi^3 + \frac{\pi^2}{2} + \frac{\pi^4}{3} - \frac{3}{2}\frac{4t^2 + \Delta_d^2}{\Delta_s^2}\left(\frac{4}{3}\pi^2 + \frac{\pi^2}{2} + \frac{81}{32}\pi^2\right) + 3\frac{4t^2 - \Delta_d^2}{\Delta_s^2}\left(\frac{4}{3}\pi^3 + \frac{\pi^2}{2}\right)\right]\right].$$

If $\frac{t}{\Delta_s} \ll 1$ and $\frac{\Delta_d}{\Delta_s} \ll 1$ then

$$K(0,0) = \frac{e^2}{mc}\left[1 - \frac{\pi^2}{6}\frac{t}{\Delta_s}\right], \qquad (93)$$

thus this is the same expression as the formulas (78) and (85) but with proviso that the hopping parameter is much lesser than the s-wave component of the gap for quartets.

9. Numerical Results

In this section, we would like to present some numerical results regarding some zero-temperature properties of the system represented by the Hamiltonian (1) with $V_{BCS} = 0$. In the beginning, let us consider the scenario with the rectangular DOS. We will show the results for the s-wave and the $d_{x^2-y^2}$-wave pairings. For the sake of clarity, let us denote the gap for the s-wave pairing case by Δ_s

whereas the gap for the $d_{x^2-y^2}$-wave pairing by Δ_d. We shall be using the coupling constant $g = 0.5$ eV whereas the bandwidth De will be equal to three values: 0.5 eV, 1 eV, 2 eV. The results are shown in Table 1.

Table 1. The numerical results for the s-wave pairing case with the rectangular DOS for the half-filled band with $n = 1$ and $\mu = \dfrac{De}{2}$. We have used the Equation (43) for numerical calculations.

De	0.5 eV	1 eV	2 eV
Δ_s	0.212729 eV	0.13786 eV	0.036643 eV
E_G/N	-0.25464 eV	-0.29734 eV	-0.50302 eV
$K(0,0)\frac{mc}{e^2}$	0.8096	0.75899	0.75017

From Table 1, one can notice that for the increasing bandwidth, the gap Δ_s clearly decreases. The ground state energy per lattice site is negative and its module increases. One needs to mention that the same results would hold in the BCS case. However, the electromagnetic kernel multiplied by $\frac{mc}{e^2}$ behaves in another way, namely, it differs from the BCS result that is equal to the unity and its value diminishes with the increasing bandwidth. This means that for a superconductor with the wider band, the Meissner effect is weaker than for one with the narrower band. Of course, the penetration depth is greater in the former case than in the latter one.

The results for the investigated system with the pure $d_{x^2-y^2}$-wave pairing symmetry are placed in Table 2.

Table 2. The numerical results for the $d_{x^2-y^2}$-wave pairing case with the rectangular DOS for the half-filled band with $n = 1$ and $\mu = \frac{De}{2}$. We have used the Equations (48)–(50) for numerical calculations.

De	0.5 eV	1 eV	2 eV
Δ_d	0.08377 eV	0.022226 eV	0.000814 eV
E_G/N	-0.14247 eV	-0.25111 eV	-0.5 eV
$K(0,0)\frac{mc}{e^2}$	0.75661	0.75012	0.75

As one can notice, the behavior of the system with this symmetry of pairing is very similar to that of the former one. The only difference is that the $d_{x^2-y^2}$-wave pairing values are lesser than those for the s-wave pairing. Especially, it is visible at looking at the gaps.

Now let us discuss the results for the one-dimensional system with the cosine dispersion relation. These results are shown in Table 3.

Table 3. The numerical results for the one-dimensional system with the s-wave pairing symmetry and the cosine dispersion relation for the half-filled band. Note that $De = 4t$ and $g = 0.5$ eV with $n = 1$ and $\mu = 0$. We have used the Equations (72)–(74) for numerical calculations.

De	0.1 eV	1 eV	2 eV
Δ_s	0.247506 eV	0.08511 eV	0.007469 eV
E_G/N	-0.12749 eV	-0.32065 eV	-0.63663 eV
$K(0,0)\frac{mc}{e^2}$	0.9199567	0.613032	0.607318

One can see the same tendencies as in the previous tables. The gap Δ decreases for the increasing bandwidth De. The ground state energy per lattice site is also negative and its module increases with the increasing bandwidth. The electromagnetic kernel decreases as well though for the bandwidths $De = 1$ eV and $De = 2$ eV this quantity is lesser than $\frac{3}{4}\frac{e^2}{mc}$. Let us add that the approximated Equation (75) has three real solutions for $g = 0.5$ eV and two hopping parameters $t = 0.01$ eV or $t = 0.025$ eV. For the former value of t one obtains: $\Delta_{s1} = -0.00981$ eV, $\Delta_{s2} = 0.010211$ eV and $\Delta_{s3} = 0.2496$ eV whereas for the latter one $\Delta_{s1} = -0.023885$ eV, $\Delta_{s2} = 0.026437$ eV and $\Delta_{s3} = 0.247448$ eV. Note that only Δ_{s3} in both cases are proper because they fulfil the inequality $\frac{2t}{\Delta} \ll 1$. Additionally, for $t = 0.025$ eV the value of Δ_{s3} is very close to that in Table 3 for $De = 0.1$ eV corresponding to $t = 0.025$ eV.

Concerning the two-dimensional system, the situation is very similar to that above. Let us start from the description of the scenario with the s-wave pairing symmetry. Table 4 contains the numerical outcomes for four values of the bandwidth $De = 8t$. As has already been stated, there are the same tendencies as in the previous cases. It is worth noting that for the bandwidth $De = 1$ eV the paramagnetic term in the electromagnetic kernel is greater than its values for the previously discussed variants of the investigated model. The penetration depth is obviously greater than corresponding values of this quantity. Moreover, the Equation (83) has three real solutions for two values of the hopping parameter $t = 0.01$ eV and $t = 0.025$ eV. They are as follows: for the former one one gets $\Delta_{s1} = -0.013768$ eV, $\Delta_{s2} = 0.014573$ eV and $\Delta_{s3} = 0.249195$ eV while for the latter $\Delta_{s1} = -0.033217$ eV, $\Delta_{s2} = 0.038433$ eV and $\Delta_{s3} = 0.244785$ eV. In this instance one obtains a very good agreement between Δ_{s3} and those values for $De = 0.08$ eV and $De = 0.2$ eV from Table 4.

Table 4. The numerical results for the two-dimensional system with the s-wave pairing symmetry and the cosine dispersion relation for the half-filled band. Note that $De = 8t$, $g = 0.5$ eV, $\mu = 0$. We have used the Equations (79), (81) and (82) with $\mu = 0$ for numerical calculations.

De	0.08 eV	0.2 eV	0.4 eV	1 eV
Δ_s	0.2492 eV	0.2451 eV	0.2319 eV	0.17263 eV
E_G/N	−0.1258 eV	−0.13015 eV	−0.1482 eV	−0.2246 eV
$K(0,0)\frac{mc}{e^2}$	0.9347	0.8431	0.7153	0.4865

In the end, we shall describe the pure $d_{x^2-y^2}$-wave pairing case. Here, the investigated quantities behave in a similar way to the corresponding s-wave pairing ones. We adopt the same dispersion relation as in the s-wave case. The results are shown in Table 5.

Table 5. The numerical results for the two-dimensional system with the $d_{x^2-y^2}$-wave pairing symmetry and the cosine dispersion relation for the half-filled band. Note that $De = 8t$, $g = 0.5$ eV, $\mu = 0$. We have used the Equations (87), (89) and (90) with $\mu = 0$ and $\Delta_s = 0$ for numerical calculations.

De	0.08 eV	0.2 eV	0.4 eV	1 eV
Δ_d	0.1997 eV	0.1933 eV	0.1831 eV	0.1551 eV
E_G/N	−0.08664 eV	−0.09957 eV	−0.12737 eV	−0.22873 eV
$K(0,0)\frac{mc}{e^2}$	0.9194	0.8075	0.6711	0.46784

In this instance one can notice that the order parameter for the s-wave pairing case is visibly greater than in the $d_{x^2-y^2}$-wave pairing one. The same assertion concerns the electromagnetic kernel. However, the ground state energy per lattice site for the latter kind of pairing and $De = 1$ eV is slightly greater than that in the former kind of pairing.

10. Conclusions

The system with the four-fermion attraction has been investigated in this paper. The original model represented by the Hamiltonian (1) comprises the BCS interaction, however, it has been neglected here for simplicity. The study has been devoted to some zero-temperature values of some quantities in this system such as the order parameter, the ground state energy per lattice site and the electromagnetic response to an external magnetic field. An essential part of the study is to get an answer to the question of how the symmetry of the pairing affects those quantities. It turns out that the influence of this factor cannot be neglected. The difference between both kinds of pairings is significant independent of what type of dispersion relation we will use. Basically, all quantities with the $d_{x^2-y^2}$-wave pairing are lesser than those with the *s*-wave pairing. Moreover, the electromagnetic kernel for the system in the frame of the tight-binding approximation (the cosine dispersion relation) is lesser than this kernel for the system with the rectangular DOS. In both of cases, the same bandwidth is used. The example with $De = 1$ eV is visible in Tables from the previous section. The interesting result is found for the system with the cosine dispersion relation. For a very small ratio $\frac{t}{\Delta_g}$ the electromagnetic kernel has approximately the same value independent of the dimension and type of pairing.

There are some open problems not undertaken in this paper. For example, properties of the system at finite temperatures are not studied here. From [18,19], it is known that there should be some differences between the conventional BCS system and the system studied in this paper. One can mention the character of the phase transition and the value of the critical temperature. Moreover, the incorporation of the BCS interaction can, to large extent, affect different properties of the system. Additionally, different types of pairings in such interactions can lead to many interesting phenomena.

Funding: This research received no external funding.

Acknowledgments: The author would like to thank R. Szczęśniak and W. Leoński for their support during the writing of this text.

Conflicts of Interest: The author declares no conflict of interest.

References

1. Maćkowiak, J.; Tarasewicz, P. A BCS-type system of interacting fermion pairs. *Phys. C* **2000**, *331*, 25–37. [CrossRef]
2. Schneider, T.; Keller, H. Extreme type II superconductors: universal properties and trends. *Int. J. Mod. Phys. B* **1993**, *8*, 487–528. [CrossRef]
3. Bunkov, Y.M.; Chen, A.S.; Cousins, D.J.; Godfrin, H. Semisuperfluidity of ^3He in Aerogel? *Phys. Rev. Lett.* **2000**, *85*, 3456. [CrossRef] [PubMed]
4. Volovik, G.E. *Exotic Properties of Superfluid ^3He*; World Scientific: Singapore, 1992.
5. Schneider, C.W.; Hammerl, G.; Logvenov, G.; Kopp, T.; Kirtley, J.R.; Hirschfeld, P.J.; Mannhart, J. Half-$h/2e$ critical current—oscillations of SQUIDs. *Europhys. Lett.* **2004**, *68*, 86. [CrossRef]
6. Aligia, A.A.; Kampf, A.P.; Mannhart, J. Quartet formation at (100)/(110) interfaces of d-wave superconductors. *Phys. Rev. Lett.* **2005**, *94*, 247004. [CrossRef]
7. Röpke, G.; Schnell, A.; Schuck, P.; Nozieres, P. Four-particle condensate in strongly coupled fermion systems. *Phys. Rev. Lett.* **1998**, *80*, 3177–3180. [CrossRef]
8. Balian, R.; Werthamer, N.R. Superconductivity with Pairs in a relative p wave. *Phys. Rev.* **1963**, *131*, 1553. [CrossRef]
9. Anderson, P.W.; Brinkman, W.F. *The Helium Liquids*; Armitage, J.G.M., Farquhar, I.E., Eds.; Academic Press: New York, NY, USA, 1975.
10. Scalapino, D.J. The case for $d\_x^2 - y^2$ pairing in the cuprate superconductors. *Phys. Rep.* **1995**, *250*, 329–365. [CrossRef]
11. Riseborough, P.S.; Schmiedeshoff, G.M.; Smith, J.L. Heavy-Fermion Superconductivity. In *Superconductivity: Novel Superconductors Vol. II*; Bennemann, K.H., Ketterson, J.B., Eds.; Springer: Berlin/Heidelberg, Germany, 2008.
12. Tarasewicz, P.; Baran, D. Extension of Fröhlich's method to 4-fermion interactions. *Phys. Rev. B* **2006**, *73*, 094524. [CrossRef]
13. Hirsch, J.E. Superconductivity from undressing. *Phys. Rev. B* **2000**, *62*, 14487. [CrossRef]
14. Szczęśniak, R. Pairing mechanism for the high-TC superconductivity: Symmetries and thermodynamic properties. *PLoS ONE* **2012**, *7*, e31873. [CrossRef] [PubMed]
15. Szczęśniak, R.; Durajski, A.P.J. Non-BCS temperature dependence of energy gap in thin film electron-doped cuprates. *Supercond. Nov. Magn.* **2016**, *29*, 1779–1786. [CrossRef]
16. Szczęśniak, R.; Durajski, A.P.; Duda, A.M. Pseudogap in the Eliashberg approach based on electron-phonon and electron-electron-phonon interaction. *Ann. Phys.* **2017**, *529*, 1600254. [CrossRef]
17. Maćkowiak, J.; Baran, D. Thermodynamics of a superconductor with 2-particle and 4-particle attraction in the strong coupling limit. *Int. J. Mod. Phys. B* **2011**, *25*, 1701–1735. [CrossRef]

18. Tarasewicz, P.; Maćkowiak, J. Thermodynamic functions of Fermi gas with quadruple BCS-type binding potential. *Phys. C* **2000**, *329*, 130–148. [CrossRef]
19. Tarasewicz, P. Thermodynamic functions of Fermi gas with quadrupole potential with $d_{x^2-y^2}$-wave pairing symmetry. *Supercond. Nov. Magn* **2004**, *17*, 431–437. [CrossRef]
20. Maćkowiak, J.; Tarasewicz, P. An extension of the Bardeen-Cooper-Schrieffer model of superconductivity. *Phys. C* **2000**, *331*, 25–37. [CrossRef]
21. Czerwonko, J. The thermodynamics of statistical spin liquid. *Phys. C* **1994**, *2377*, 235–240. [CrossRef]
22. Czerwonko, J. The thermodynamics of statistical spin liquid with the S- and D-pairing. *Mol. Phys. Rep.* **1995**, *12*, 79.
23. Abrikosov, A.A. *Fundamentals of the Theory of Metals*; North Holland: Amsterdam, The Netherlands, 1978.
24. Tarasewicz, P. On Meissner effect in a superconductor with 4-fermion attraction. *Eur. Phys. J. B* **2004**, *41*, 185. [CrossRef]

© 2019 by the author. Licensee MDPI, Basel, Switzerland. This article is an open access article distributed under the terms and conditions of the Creative Commons Attribution (CC BY) license (http://creativecommons.org/licenses/by/4.0/).

MDPI
St. Alban-Anlage 66
4052 Basel
Switzerland
Tel. +41 61 683 77 34
Fax +41 61 302 89 18
www.mdpi.com

Symmetry Editorial Office
E-mail: symmetry@mdpi.com
www.mdpi.com/journal/symmetry

www.ingramcontent.com/pod-product-compliance
Lightning Source LLC
LaVergne TN
LVHW070042120526
838202LV00101B/386